W0051106

SOVIET RESEARCH IN NEW SEMICONDUCTOR MATERIALS

ISSLEDOVANIYA PO POLUPROVODNIKAM:

NOVYE POLUPROVODNIKOVYE MATERIALY

ИССЛЕДОВАНИЯ ПО ПОЛУПРОВОДНИКАМ:

НОВЫЕ ПОЛУПРОВОДНИКОВЫЕ МАТЕРИАЛЫ

Soviet Research in
NEW SEMICONDUCTOR MATERIALS

Edited by

D. N. Nasledov

and

N. A. Goryunova

Authorized translation from the Russian by
A. Tybulewicz, B.Sc., A.Inst.P., M.I.Inf.Sc., F.I.L.

Springer Science+Business Media, LLC

1965

Editorial Board:
D. N. Nasledov, N. A. Goryunova, D. V. Gitsu,
V. N. Lange, and S. I. Radautsan

The original Russian text was published by the
Soviet State Publishing House, "Kartya Moldovenyaské"
in Kishinev, 1964.

ISBN 978-1-4899-4946-2   ISBN 978-1-4899-4944-8 (eBook)
DOI 10.1007/978-1-4899-4944-8

Library of Congress Catalog Card Number 65-11956

© 1965 Springer Science+Business Media New York
Originally published by Consultants Bureau Enterprises, Inc in 1965.
Softcover reprint of the hardcover 1st edition 1965
All rights reserved

No part of this publication may be reproduced in any
form without written permission from the publisher

CONTENTS

DIAMOND-LIKE DEFECT SEMICONDUCTORS

N. A. Goryunova and S. I. Radautsan

Among the large number of complex semiconducting materials with diamond-like structure that have been investigated recently, the defect compounds and alloys based on them are of particular interest [1].

In 1949, Hahn and Klingler [2] established that some compounds of the $A_2^{III}B_3^{VI}$ type, namely Ga_2S_3, Ga_2Se_3, Ga_2Te_3, and In_2Te_3, crystallize in the zinc blende structure with one-third of the cation sites vacant due to the absence of gallium or indium atoms.

Even earlier, Ketelaar [3] found that the ternary compounds Ag_2HgI_4 and Cu_2HgI_4 crystallize in tetrahedral lattices with defects in the cation part.

It is known that the structure of zinc blende in normal nondefect compounds is associated with the existence of tetrahedral bonds involving electron pairs. The work cited above suggested that tetrahedral bonds may not be retained in defect compounds: the number of electrons is sufficient to form such bonds but some of these bonds would have to start from vacant sites. Some ways of resolving this difficulty were suggested by studies of the isomorphism of these compounds.

Experiments on the possibility of the formation of solid solutions, in a wide range of concentrations, between compounds $A_2^{III}B_3^{VI}$, on the one hand, and $A^{III}B^V$ or $A^{II}B^{VI}$, on the other, suggested that these compounds are not only similar in structure but also in their type of binding (nature of the electron density distribution) and, moreover, that they should have semiconducting properties [4-6]. Further investigations of the electrical properties of these defect compounds and their solid solutions have indeed shown that all of them are semiconductors [7-10].

Although several compounds of $A_2^{III}B_3^{VI}$ do not have sphalerite structure at room temperature (In_2S_3, In_2Se_3, Al_2Te_3, etc.), it has been found that when they are fused with $A^{III}B^V$ or $A^{II}B^{VI}$ compounds, solid solutions are formed in a wide range of concentrations, the structure of these solutions being that of zinc blende, sometimes ordered [5, 11-14].

The work just cited has explained many characteristic features of these systems and has thus attracted the attention of other investigators. In the last 2-3 years, physico-chemical and some electrical and optical properties of the alloys with various degrees of defect structure have been investigated in detail.

The present paper reviews the investigations of binary defect compounds, as well as the solid solutions and complex alloys based on these compounds.

SYNTHESIS AND INVESTIGATION METHODS

The majority of diamond-like defect substances has been prepared by fusing very pure elements in evacuated quartz ampoules at temperatures above the melting point of the corresponding binary compound [15]. Many investigators washed and then filled the ampoules with an inert gas, particularly with spectroscopically pure argon.

Since, at high temperatures, the vapor pressure of many of the elements involved was very high, ampoules frequently exploded during synthesis. The use of vibration mixing [16, 17] and melting in a double-temperature furnace [18] eliminated these explosions.

Some workers prepared samples by the pressing and prolonged heating of a mixture of binary compounds [11]. In this way, partly fused and partly fired materials were obtained, which were unsuitable for studies of the electrical properties, microstructure, and microhardness.

The substances were purified by zone recrystallization [19, 20]. Single-crystal samples were obtained by the Bridgman method [19]. The method of chemical transport reactions [21] was found to be a very promising one for growing single crystals of defect compounds, particularly refractory compounds.

To homogenize complex defect semiconducting materials, the following methods have been used:

1) prolonged annealing of powdered or bulk samples [15];

2) annealing under pressure [22];

3) zone leveling at low molten-zone velocities (0.2-12 mm/hour) [10];

4) directional crystallization of an ingot and slow cooling [23].

Each of these homogenization methods has its advantages and disadvantages. Therefore, to obtain uniform alloys in a system it is necessary to select the most effective method for the particular material.

Diamond-like defect semiconductors, as mentioned above, have been investigated only in the last few years. Therefore, the majority of the papers published so far deal only with the structures of these phases and with the possibility of, and conditions for, solid-solution formation.

For a relatively small number of systems, the phase diagrams have been investigated. For some alloys, the temperature dependences of the electrical conductivity, thermoelectric power, Hall coefficient, thermal conductivity, and optical properties have been studied.

A. BINARY DEFECT COMPOUNDS

$A_2^{III}B_3^{VI}$ compounds with a defect zinc blende structure occupy, in respect of their chemical composition, an intermediate position in the isoelectronic series of the diamond group, between $A^{III}B^V$ and $A^{II}B^{VI}$ compounds. However, almost all the properties of $A_2^{III}B_3^{VI}$ compounds, except the forbidden band width and microhardness, do not have values intermediate between the corresponding values for neighboring non-defect compounds in a given isoelectronic series, i.e., in a series of substances having the same number of inner electrons.

Table 1 lists some properties of the isoelectronic series of germanium and of gray tin [9, 10, 15, 24, 25]. This table shows that the compounds Ga_2Se_3, Ga_2Te_3, and In_2Te_3 differ from other terms of the isoelectric series by a somewhat lower value of the lattice constant, very low mobility and thermal conductivity. This is due to an "intrinsic" defect structure. The lattice contracts because the atoms surrounding cation vacancies have excess energy due to the absence of one nearest neighbor [24]. The cation vacancies distort the potential distribution in the lattice and the lattice contraction decreases the degree of anharmonicity of the forces between, atoms, thereby reducing the phonon mean free path and intensifying electron scattering.

The ionic nature of the binding becomes stronger in the series (arranged according to chemical composition) away from the element A^{IV}. This accounts for the monotonic variation of the forbidden band width and microhardness in these series.

Table 2 lists the structure of $A_2^{III}B_3^{VI}$ crystals [26-30]. This table shows that all aluminum and gallium chalcogenides crystallize in tetrahedral structures. The only exception is indium: only In_2Te_3 has the sphalerite structure ordered at room temperature (the α phase) and disordered above 450°C (the β phase) [28]. Recent electron-diffraction studies of indium selenide have shown that one of its high-termperature modifications (the γ phase) has cubic structure [27]. X-ray diffraction studies, carried out at the Institute of Physics and Mathematics of the Moldavian Academy of Sciences, have confirmed Semiletov's results [27]. A study of In_2Se_3

TABLE 1. Variation of the Properties in Several Isoelectric Series

Property \ Substance	A^{IV}	$A^{III} B^V$	$A^{III}_2 B^{VI}_3$	$A^{II} B^{VI}$	$A^I B^{VII}$	Number of inner electrons
	Ge	GaAs	Ga₂Se₃	ZnSe	CuBr	28
Lattice parameter, A	5.65	5.63	5.43	5.66	5.68	
Forbidden band width ΔE, eV	0.78	1.35	1.98	2.8	2.9	
Electron mobility, cm²·V⁻¹·sec⁻¹	3.900	6.000	10	200	30	
Microhardness, kg/mm²	1000	720	310	135	21	
$\varkappa_{ph} \times 10^3$, cal·cm⁻¹·sec⁻¹ deg⁻¹	140	125	1.22	33	—	
	Ge Sn	GaSb	Ga₂Te₃	ZnTe	CuI	28 · 46
Lattice parameter, A	—	6.09	5.88	6.08	6.05	
Forbidden band width ΔE, eV	—	0.68	1.2	2.2	—	
Electron mobility, cm²·V⁻¹·sec⁻¹	—	5000	50	300	—	
Microhardness, kg/mm²	—	420	240	100	19	
$\varkappa_{ph} \times 10^3$, cal·cm⁻¹·sec⁻¹ deg⁻¹	—	105	1.12	34	—	
	Sn	InSb	In₂Te₃	CdTe	AgI	46
Lattice parameter, A	6.49	6.46	6.16	6.46	6.47	
Forbidden band width ΔE, eV	0.08	0.18	1.12	1.5	2.8	
Electron mobility, cm²·V⁻¹·sec⁻¹	2.000	100.000	∼10	450	30	
Microhardness, kg/mm²	—	220	200	60	7	
$\varkappa_{ph} \times 10^3$, cal·cm⁻¹·sec⁻¹·deg⁻¹	—	37	1.66	15	—	

TABLE 2. Structures of $A_2^{III}B_3^{VI}$ Compounds

	O_3	S_3	Se_3	Te_3
B_2	Hexagon. cubic	no data	no data	no data
Al_2	$\alpha_1 - Al_2O_3$ $\gamma_1 - Fe_2O_3$	$\alpha_1 - W_x$ $\beta_1 - W$ $\gamma_1 - Al_2O_3$	W_x W	W
Ga_2	$\alpha_1 - Al_2O_3$ $\beta_1 -$ monoclinic, stable	$\alpha - W_x$ $\beta - W > 650$ $\gamma - S$	S	S
In_2	Tl_2O_3	$\alpha_1\gamma' - Al_2O_3$ β - spinel, stable	α_1 - hexagon β - hexagon. γ - cubic	$\alpha - S_x$ low temp. $\beta - S$

Notation: W - defect wurtzite structure
W_x - ditto, with ordered vacancies
S - defect sphalerite structure
S_x - ditto, with ordered vacancies

by means of a camera constructed by V. G. Kuznetsov (working at 600°C) has shown the presence of lines representing the structure of zinc blende.

Table 2 shows that many defect compounds have several modifications existing at various temperatures.

The phase diagrams of $A^{III}B^{VI}$ systems show, as a rule, several compounds. By way of example, Figs. 1 and 2 show the phase diagrams for the gallium—tellurium and indium—tellurium systems [31]. The former system has three compounds: GaTe, Ga_2Te_3, and $GaTe_3$; the latter has four: In_2Te, InTe, In_2Te_3, and In_2Te_5. The existence of the compound In_4Te_7 has been recently reported [32].

It is worth pointing out that in $A^{III}B^{V}$ and $A^{II}B^{VI}$ systems there is usually only one compound [1].

In $A^{III}B^{VI}$ systems, the melting points of sulfides and selenides differ from those of tellurides. Figures 1 and 2 show that the melting point of tellurides of the AB type is greater than the melting point of tellurides of the $A_2^{III}B_3^{VI}$ type. For gallium and indium sulfides and selenides, the position is reversed. This may be important in the determination of the nature of the interaction in more complex systems, where the formation of refractory substances is favored by energy considerations.

Some properties of the best-known $A_2^{III}B_3^{VI}$ compounds are given in Table 3.

Indium telluride, In_2Te_3, has been investigated in some detail in its two modifications: the low-temperature ordered α phase and the high-temperature (> 450°C) disordered β phase.

Figures 3 and 4 give the temperature dependences of the electrical conductivity, the Hall coefficient, and the thermoelectric power in the intrinsic conduction region of In_2Te_3 samples ordered to various degrees [24]. It has been found that the intrinsic conduction region of In_2Te_3 is observed even at -100°C.

The forbidden band width of α-In_2Te_3, calculated from $\log RT^{3/2} = f(1/T)$ in Fig. 3, is 1.12 ± 0.05 eV. The carrier mobility in the intrinsic conduction region is independent of temperature and the effective masses are close to the free-electron mass ($m_n \approx 0.7m_0$, $m_p \approx 1.1m_0$). The thermal conductivity of the disordered β-In_2Te_3 (Table 3) is considerably lower and, in contrast to the α phase, is almost independent of temperature [34]. This is due to the phonon mean free path being comparable with the interatomic spacing in these crystals, which leads to strong phonon scattering on lattice vacancies. Therefore, we conclude that the phonon mean free path may be a sensitive indicator of the process of ordering in the cation sublattice.

4

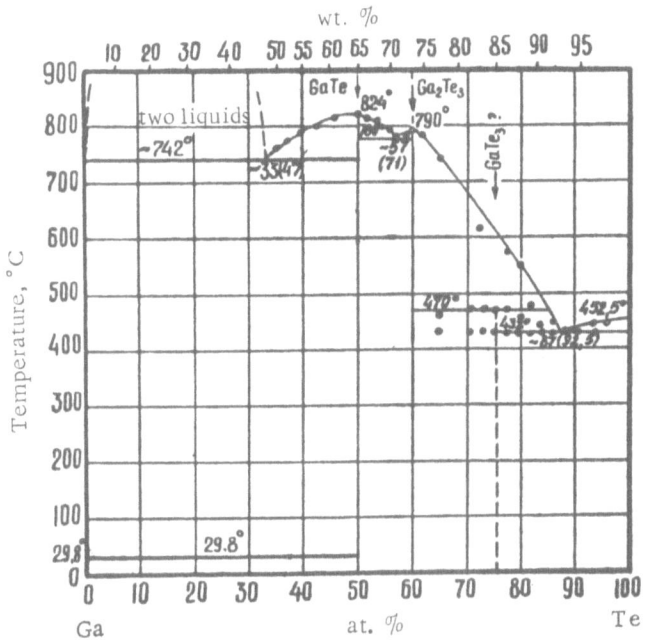

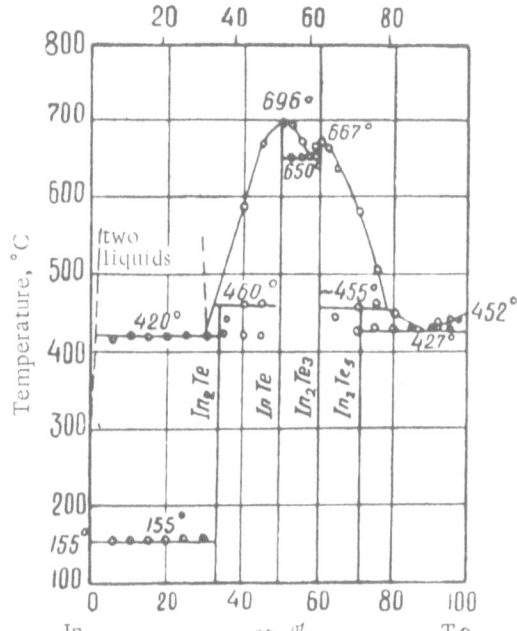

Fig. 1. Phase diagram of the Ga—Te System [31]. Fig. 2. Phase Diagram of the In—Te System [31].

The alloying of indium telluride with various impurities up to 1 at. % does not lead to a marked change in its electrical properties (the exceptions are bismuth and iodine impurities). Following the theoretical treatment of Fisher [35], which explained the low electrical activity of impurities in amorphous and glassy semiconductors, we may assume that defect semiconductors are an intermediate link between normal crystals and amorphous substances.

Investigations of the influence of small amounts of copper (10^{-6}-10^{-1} at. %) on the properties of Ga_2Te_3 have shown that the lattice constant rises by 2.7% and the forbidden band becomes narrower [36]. These results have not yet been explained. Nasledov and Feltyn'sh investigated the electrical properties of Ga_2Se_3 [9] and found that this compound is p-type up to 1000°K. They suggested that the hole mobility in gallium selenide is higher than the electron mobility. Such a result distinguishes this compound from its nondefect analogs of the $A^{III}B^V$ type, in which the mobility of electrons is higher than that of holes. On the other hand, it has recently been shown that in the complex glassy alloy $Tl_2Se-As_2Te_3$ the hole mobility is one order of magnitude greater than the electron mobility [37].

The present authors [8, 29] have investigated the photoelectric properties of Ga_2Se_3, Ga_2Te_3, and In_2Te_3. The forbidded band widths calculated from the long-wavelength edge of the photoconductivity are in good agreement with the electrical and optical data.

Investigations of the absorption spectra of β-Ga_2S_3 and Ga_2Se_3 at 77.3 and 290°K have shown no line structure of the absorption edge [38] which is associated with the appearance of exciton states in crystals. This result has been explained as being due to the influence of a large number of defects in $A_2^{III}B_3^{VI}$ compounds, which impede exciton formation and their migration across the lattice. The forbidden band widths at room termperature are close to the values obtained from the photoelectric measurements [8] and are, respectively, 2.5-2.7 eV for β-Ga_2S_3 and 1.9-2.1 eV for Ga_2Se_3.

The properties of films of Al_2Se_3, Ga_2Te_3, and In—Se alloys of variable composition have also been investigated [39, 40]. The results are in good agreement with the data obtained for bulk samples.

B. $A_2^{III}B_3^{VI} - A_2^{III}B_3^{VI}$ SYSTEMS

We shall now consider pseudobinary systems of $A_2^{III}B_3^{VI}$ compounds, in which isovalent substitution has been observed.

5

TABLE 3. Some Properties of $A_2^{III} B_3^{VI}$ -Type Compounds

Compound	Color	Structure	Lattice parameter, Å	Density, g/cm³	Microhardness under 20 g, kg/mm²	Thermal conductivity at 300°K, $\varkappa_{ph} \times 10^3$ cal·cm⁻¹·sec⁻¹·deg⁻¹	ΔE_{opt} at 300°K, eV	μ at 300°K, cm²·V⁻¹·sec⁻¹	Melting point, °C
α - Al₂S₃	white	α – hexag.	a = 6.423 c = 17.83	2.5			4.1		1.130
		β – wurtzite	a = 3.58 c = 5.83	2.3					
Al₂Se₃	white	wurtzite	a = 3.89 c = 6.30	3.9			3.1		980
Al₂Te₃	grayish yellow	wurtzite	a = 4.08 c = 6.94	4.5			2.2		895
Ga₂S₃	light yellow	α – sphalerite, low-temp.	5.181				2.5		1.250
		β – wurtzite, high-temp.	a = 3.678 c = 6.016	3.63	500				
Ga₂Se₃	red	sphalerite	5.429	4.92	316	1.22	1.9	10	1.020
Ga₂Te₃	black	sphalerite cubic,	5.886	5.57	237	1.12	1.35	50	790
In₂S₃	red	low-temp. spinel > 300°C	a = 5.37	4.63	280		1.2		1.050
			a = 10.74						
In₂Se₃	black	α₁ – hexag.	a = 4.02 c = 19.2	5.48	30 – 50	2.5	1.02	125	890
		α – sphalerite	18.49	5.79		2.68		2 – 10	
In₂Te₃	black	β – sphalerite	6.158	5.73	166	1.66	1.026	50	667

The data in the table are taken from [1, 26, 30, 33, 39].

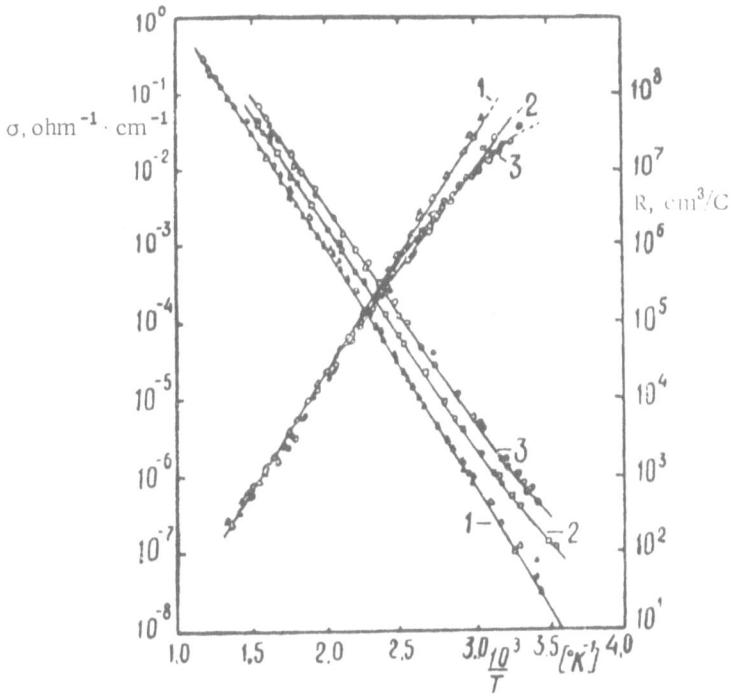

Fig. 3. Temperature dependence of the electrical conductivity and Hall coefficient of three samples of the compound α-In_2Te_3 [24].

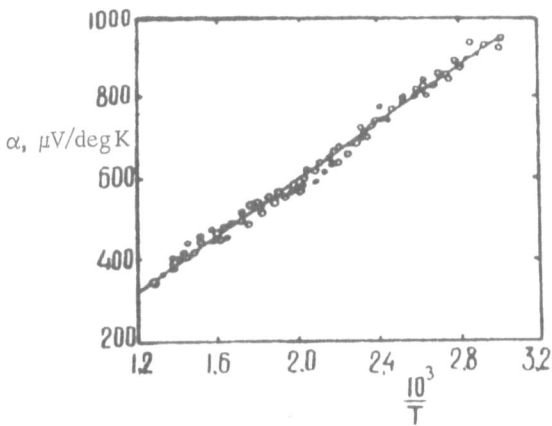

Fig. 4. Temperature dependence of the thermoelectric power of the compound α-In_2Te_3 [24].

These systems have been mostly investigated from the point of view of the possibility of the formation of solid solutions. The mutual solubility of defect $A_2^{III}B_3^{VI}$ compounds (gallium and indium tellurides) was first discovered by Goryunova in 1955 [6]. Table 4 lists the results of a study of four systems, with an indication of the solubility limits.

We shall consider in greater detail the solid solutions of defect compounds, both with anion and cation substitution.

Ga_2Se_3–Ga_2Te_3 System. This was investigated by Woolley and Smith [41] and by Grigor'eva [42]. In the 40-100 mol. % Ga_2Te_3 region, equilibrium alloys with the zinc blende structure are formed.

The compositions with 5-35 mol. % Ga_2Te_3 have exhibited broad lines in their Debye powder patterns, which are characteristic of the disordered state. After four-week's annealing of the alloys with 25-35 mol. % Ga_2Te_3, ordering was detected with a unit tetragonal cell having dimensions of $a_0 \times a_0 \times 2a_0$ (where a_0 is the lattice parameter of ZnS). The compositions with 5, 10, and 15 mol. % Ga_2Te_3 exhibited anneal-broadening of the Debye powder pattern lines with odd Miller indices, but the lines with even indices remained sharp [41]. The investigators concluded that these alloys were partly ordered. Some departure from Vegard's law was observed in this range of concentrations.

Ga_2Te_3–In_2Te_3 System. This was investigated by Goryunova [6], Grigor'eva [43], Woolley and Smith [41].

TABLE 4. Interaction in the $A_3^{III}B_3^{VI} - A_2^{III}B_3^{VI}$ Systems

System	Solid solution range	References
$Ga_2Se_3 - Ga_3Te_3$	Solid solutions with ordering near Ga_2Se_3	41,42
$Ga_2Te_3 - In_2Te_3$	Continuous solid solutions	6,41,43
$Ga_2Se_3 - In_2Se_3$	*45 mol. % with ZnS structure on the Ga_2Se_3 side	44
$In_2Se_3 - In_2Te_3$	*25 mol. % with ZnS structure on the In_2Te_3 side	45

*Limited solid solutions.

Continuous solid solutions with the zinc blende structure were obtained; the lattice constant varied from 5.87 to 6.16 A. Alloy ordering was not observed. Woolley and Smith [41] showed that the phase diagram, plotted from thermal analysis data, is complex.

$Ga_2Se_3 - In_2Se_3$ System. Ageeva, Vaipolin, and Grigor'eva [44] detected the formation of homogeneous phases over a wide range of concentrations with a morphotropic transition between structures. When the indium selenide concentration in alloys is increased, the sphalerite structure changes into the wurtzite structure in the composition region 75-70 mol. % Ga_2Se_3, and then into another structure with a hexagonal lattice.

Utilizing some features of the unit cell's shape, a comparison of the reduced cell parameters of these alloys was made and some regularities in dependence on composition were found.

$In_2Te_3 - In_2Se_3$ System. A preliminary x-ray diffraction study of alloys of this system [45] showed that near indium telluride there is a solubility region of not less than 25 mol. %. Solid solutions have the sphalerite structure with the lattice parameter decreasing from 6.16 to 6.05 A. Thermal and microstructure studies have confirmed these results. The investigators suggested the existence of a second homogeneous region but having a different structure. It is possible that this system too exhibits a morphotropic transition on increase of the indium selenide concentration.

As pointed out earlier, defect compounds of $A_2^{III}B_3^{VI}$ type may interact with nondefect compounds of $A^{III}B^V$ or $A^{II}B^{VI}$ type, forming broad homogeneous systems. We shall now consider such solid solutions, formed by heterovalent substitution.

C. $A^{III}B^V - A_2^{III}B_3^{VI}$ SYSTEMS

Among systems of this type, the best known are alloys based on compounds of relatively low melting points, i.e., based on indium and gallium compounds (Table 3). Alloys based on boron and aluminum compounds have not been investigated much because of the difficulties of high-temperature synthesis and because they rapidly decompose in the atmosphere.

Table 5 lists briefly the characteristics of the interaction of such compounds in some of the systems.

1. Systems Based on Boron

Of all possible systems of this type, only the $B_3P_3 - B_2Se_3$ alloys have been investigated. The results of x-ray diffraction analysis have shown the possibility of solid solution formation. The composition 1:1 consists of a single phase and has the pure zinc blende structure. The lattice parameter varies from 4.53 A for boron phosphide to 4.57 A for the 50:50% composition. Some compositions closer to boron selenide exhibit additional lines, apart from the ZnS structure. We may assume then that in this system the solubility extends over a range of not less than 50 equimol. % on the boron phosphide side.

TABLE 5. Interaction in the $A_2^{III}B_3^V - A_2^{III}B_3^{VI}$ Systems

System	Solid solution range	References
$B_3P_3 - B_2Se_3$	* ≈50 equimol. % on the BP side	
$Al_3Sb_3 - Al_2Te_3$	* ≈45 equimol. % on the AlSb side	46
$Ga_3P_3 - Ga_2S_3$	* ≈70 equimol. % on the GaP side	47
$Ga_3P_3 - Ga_2Se_3$	Continuous solid solutions	48
$Ga_3P_3 - Ga_2Te_3$	* ≈10 equimol. % on the Ga_2Te_3 side	
$Ga_3As_3 - Ga_2S_3$	* ≈55 equimol. % on the Ga_2S_3 side	49
$Ga_3As_3 - Ga_2Se_3$	Continuous solid solutions	6,50,41
$Ga_3As_3 - Ga_2Te_3$	* ≈60 equimol. % on the Ga_2Te_3 side	41
$Ga_3Sb_3 - Ga_2Se_3$	** Complex interaction	41
$Ga_3Sb_3 - Ga_2Te_3$	** Complex interaction	41
$In_3P_3 - In_2S_3$	** Complex interaction	52
$In_3P_3 - In_2Se_3$	* ≈75 equimol. % on the InP side	52−56,58
$In_3P_3 - In_2Te_3$	** Complex interaction	52
$In_3As_3 - In_2S_3$	* ≈50 equimol. % on the InAs side	52
$In_3As_3 - In_2Se_3$	* ≈90 equimol. % on the InAs side	10,12,56−59
$In_3As_3 - In_2Te_3$	Continuous solid solutions	10,41,60−63,58
$In_3Sb_3 - In_2S_3$	** Complex interaction	52
$In_3Sb_3 - In_2Se_3$	* ≈2.5 equimol. % on the InSb side	52,59
$In_3Sb_3 - In_2Te_3$	* ≈15 equimol. % on the InSb side	64,65

*Limited solid solutions

**Narrow solubility region possible near initial binary compounds

Note: in Table 5 and later, the compounds $A^{III}B^V$ are referred to as $A_3^{III}B_3^V$ to retain the equimolecular relationship in the interaction with $A_2^{III}B_3^{VI}$ compounds.

2. Systems Based on Aluminum

We know of only one study of a system of this type, namely $Al_3Sb_3-Al_2Te_3$ [46]. The alloys were prepared from the initial components or the corresponding alloys in corundum crucibles in an argon atmosphere. In air these alloys rapidly corrode. The general nature of the melting diagram shows a wide solid-solution region right up to 45 wt. % on the aluminum antimonide side. The determination of the temperature dependence of the electrical conductivity by the rotating magnetic field method has made it possible to determine more accurately the solidus line in the melting diagram. In contrast to $A^{III}B^V$ compounds [66], which exhibit a discontinuous change from solid into liquid state, in the solid solutions considered here the process occurs over a certain range of temperatures defining the beginning and end of melting. In the region of solubility of Al_2Te_3 in AlSb, an almost linear reduction of the lattice parameter and an increase of the density are observed on increase of the aluminum telluride concentration.

According to the thermoelectric power measurements, the alloys are n-type up to 20 wt. %, while other compositions, including aluminum antimonide, are p-type.

Measurements of the current–voltage characteristics of solid solutions showed that on increase of the aluminum telluride content the reverse voltage increased from 3-5 V for AlSb to 18 V for an alloy with 12.8 wt. % Al_2Te_3.

3. Systems Based on Gallium

Table 5 shows that quite a few heterovalent systems based on gallium have been investigated. We shall consider these systems in three separate groups: phosphides, arsenides, and antimonides.

In the phosphide group, we shall consider three systems: $Ga_3P_3-Ga_2S_3$, $Ga_3P_3-Ga_2Se_3$, and $Ga_3P_3-Ga_2Te_3$.

$Ga_3P_3-Ga_2S_3$ System. This system was investigated recently [47]. X-ray diffraction methods have shown the possibility of the formation of solid solutions with structures of the zinc blende type in the range up to 70 equimol. % on the gallium phosphide side. A preliminary study of the electrical properties has shown that the samples have high resistivity and the temperature dependence of their electrical conductivity is of the semiconducting type.

$Ga_3P_3-Ga_2Se_3$ System. In this system, the solubility was found to extend over the whole range of concentrations [48]. All alloys have the sphalerite structure with the lattice parameter decreasing from 5.436 A for gallium phosphide to 5.422 A for gallium selenide. Some alloys exhibit weak superstructure lines in Debye powder patterns, which may indicate a certain degree of ordering. Microstructure analysis and microhardness measurements confirm well the x-ray diffraction data.

$Ga_3P_3-Ga_2Te_3$ System. Preliminary investigations of alloys of this system, carried out at the Moldavian Academy of Sciences, showed that gallium phosphide may be soluble in gallium telluride at least in the range of 10 equimol. %.

In the arsenide group, the following defect systems have been investigated: $Ga_3As_3-Ga_2S_3$, $Ga_3As_3-Ga_2Se_3$, and $Ga_3As_3-Ga_2Te_3$.

$Ga_3As_3-Ga_2S_3$ System. Two solid solution regions with the sphalerite structure were detected: in the range 56 equimol. % on the gallium sulfide side and 8 equimol. % on the gallium arsenide side [49]. Within the homogeneous region (100-56 equimol. % Ga_2S_3), the microhardness exhibited a maximum (1070 kg/mm^2) at 43% Ga_2S_3.

There is a report of the photosensitivity of these alloys in the visible part of the spectrum. The investigators suggest that using more efficient homogenization, a continuous series of solid solutions should be obtained in these systems.

$Ga_3As_3-Ga_2Se_3$ System. This system was the first to be investigated and is the best known of $A^{III}B^V - A_2^{III}B_3^{VI}$ systems. It was first investigated in 1955 by Goryunova [6]. Later, Goryunova and Grigor'eva reported [50] wide regions of solubility in alloys of this system, but some of them were in a nonequilibrium

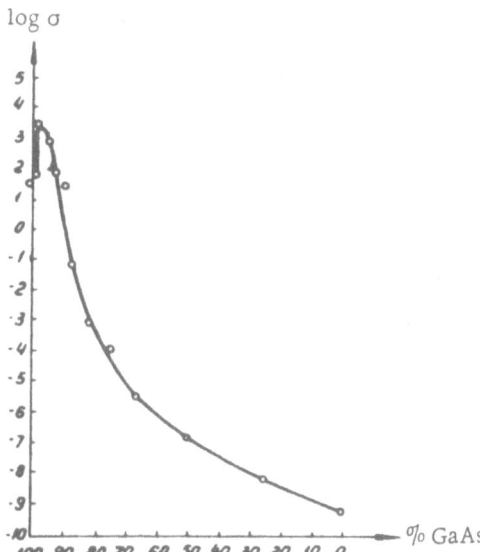

log σ

% GaAs

Fig. 5. Variation of the electrical conductivity of the GaAs−Ga$_2$Se$_3$ system with its composition [9].

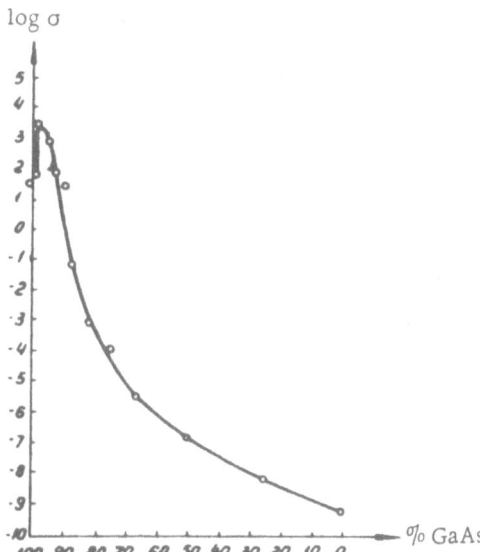

TABLE 6. Variation of the Forbidden Band Width in the $(GaAs)_{3x}-(Ga_2Se_3)_{1-x}$ System

Composition: x	ΔE, eV
1	1.35
0.7	1.4
0.5	1.6
0.25	1.7
0	1.98

state as shown by the x-ray diffraction data. Microstructure and thermal analysis data showed that the system forms a continuous series of solid solutions. The same conclusion was reached by Wolley and Smith in 1958 [41]; they annealed alloys of this system for one or two months. Over the whole range of concentrations the sphalerite structure was found with the lattice periods deviating somewhat from Vegard's law in a region close to Ga$_2$Se$_3$. Alloys of this system have a photosensitivity whose maximum moves gradually with composition from 0.7μ for Ga$_2$Se$_3$ to 1.2μ for GaAs.

The electrical properties of alloys of the Ga$_3$As$_3$−Ga$_2$Se$_3$ system were investigated by Nasledov and Feltyn'sh [9]. Figure 5 shows that the introduction of small amounts of gallium selenide into gallium arsenide sharply raises the electrical conductivity, which reaches a maximum and then decreases monotonically. This is because small amounts of selenium act as a donor impurity at gallium arsenide lattice sites, increasing the conductivity. Further increase of the concentration of the poorly conducting gallium selenide produces a substitutional solid solution and this reduces the electrical conductivity. Table 6 lists the results of measurements of the forbidden band width, determined from the temperature dependence of the electrical conductivity and the Hall effect. It is evident from Table 6 that the dependence of the value ΔE on composition departs slightly from linearity.

Measurements of the thermoelectric power of this system have shown that in the region of the electrical conductivity maximum there is some reduction of the thermoelectric power, which then increases and reaches its maximum at the composition x = 0.5, and then it again decreases, changing its sign at x = 0.12. This is in agreement with n-type conduction for gallium arsenide and p-type for gallium selenide.

Ga$_3$As$_3$−Ga$_2$Te$_3$ System. Solid solutions in this system were detected by Woolley and Smith [41] in the range 65 equimol. % on the Ga$_2$Te$_3$ side. The lattice periods of the sphalerite structure in the solubility region vary with composition following a curve lying under the Vegard line. No ordering was found in alloys of this system.

Very little work has been done on systems based on gallium arsenide. The Debye powder patterns of the middle range of compositions of the systems Ga$_3$Sb$_3$−Ga$_2$Se$_3$, and Ga$_3$Sb$_3$−Ga$_2$Te$_3$ have a large number of lines [41] which do not correspond to the zinc blende structure. Goryunova [15] showed that solid solutions may form in the GaSb−Ga$_2$Te$_3$ system.

A study of the ternary system gallium−antimony−tellurium [67] revealed a solid-solution region on the gallium antimonide side, extending up to 86.4 wt. % along the GaSb−GaTe tie-line.

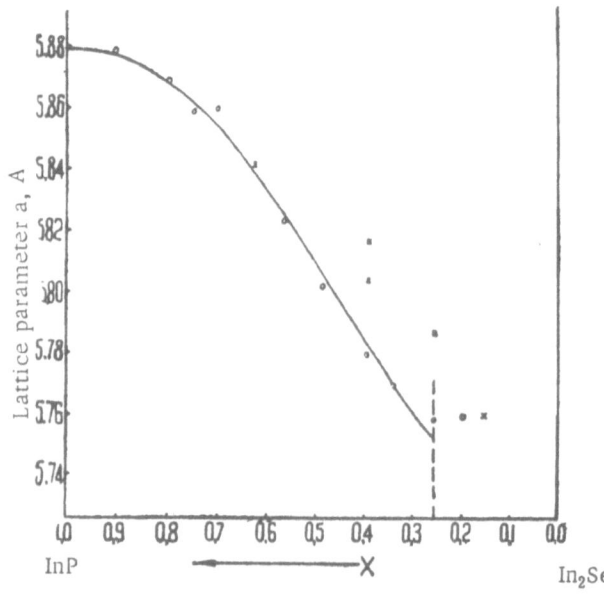
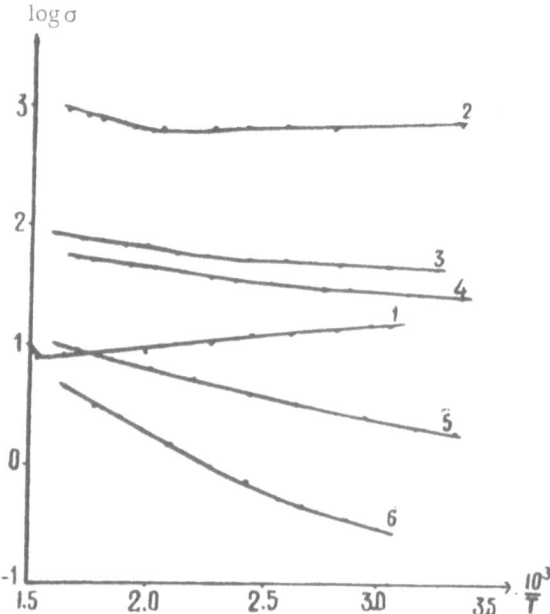

Fig. 6. Variation of the lattice parameter in the $(InP)_{3x}-(In_2Se_3)_{1-x}$ system [52, 55, 58]: O - according to [55]; ×-according to [58].

Fig. 7. Temperature dependence of the electrical conductivity of the $InP-In_2Se_3$ system [55].

4. Systems Based on Indium

Systems based on indium compounds have been studied most. This is because of important semiconducting properties of such compounds as indium arsenide or indium antimonide, which have led to the wide use of these compounds in technology, and due to the relatively low melting points of phases containing indium, which ease the technological problems of synthesis and homogenization of these alloys.

We shall consider ternary systems based on indium, grouping them as in preceding cases, on the basis of $A^{III}B^V$ compounds.

a) Systems Based on Indium Phosphide. Of three systems based on InP (Table 5), only indium phosphide—selenides form substitutional solid solutions over a wide range of concentrations [52-55, 58]. In the $In_3P_3-In_2S_3$, and $In_3P_3-In_2Te_3$ systems solid solutions are formed only in a narrow range of concentrations near the initial binary compounds [52].

For the $In_3P_3-In_2Se_3$ system, a combined study of several physico-chemical properties has established solubility in the concentration range up to 75 equimol. % [55]. Hahn and Thiele [58] used x-ray diffraction analysis to establish a solubility region of the order of 85.7 equimol. % (up to 33.3 mol. % InP). Figure 6 shows the variation of the lattice constant with composition, according to the work reported in [55] and [58]. It is evident from Fig. 6 that the numerical results of the two studies differ somewhat, the lattice constants reported by Hahn and Thiele being somewhat high. It is very likely that this is the result of insufficient homogeneity since Hahn and Thiele did not melt but fired the components at 650°C and followed this by prolonged annealing. This conclusion is supported also by the fact that for 40 equimol. % InP Hahn and Thiele obtained two values of the lattice constant, 5.80_4 A and 5.81_6 A, corresponding to two phases containing different amounts of indium selenide. Approximately the same results have been obtained by the present authors in the early investigations of this system when the optimal synthesis conditions have not yet been determined. Figure 6 shows clearly that near InP, approximately up to alloys of composition x = 0.75, the lattice constant varies little, but it deviates from Vegard's law.

The results of thermal and microstructural analyses [52, 55] confirm the presence of solid solutions in the range up to 75 equimol. %.

12

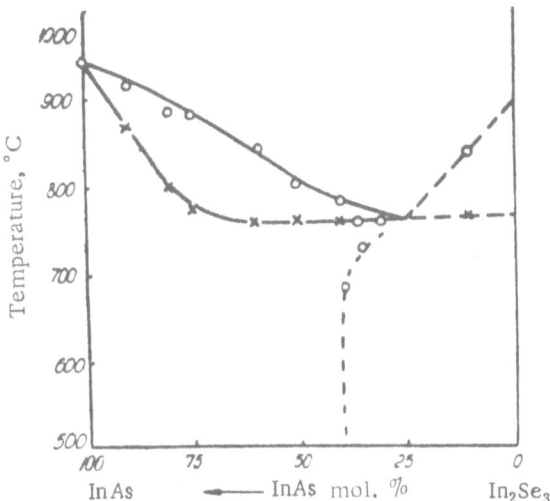

Fig. 8. Phase diagram of the InAs−In₂Se₃ system [56].

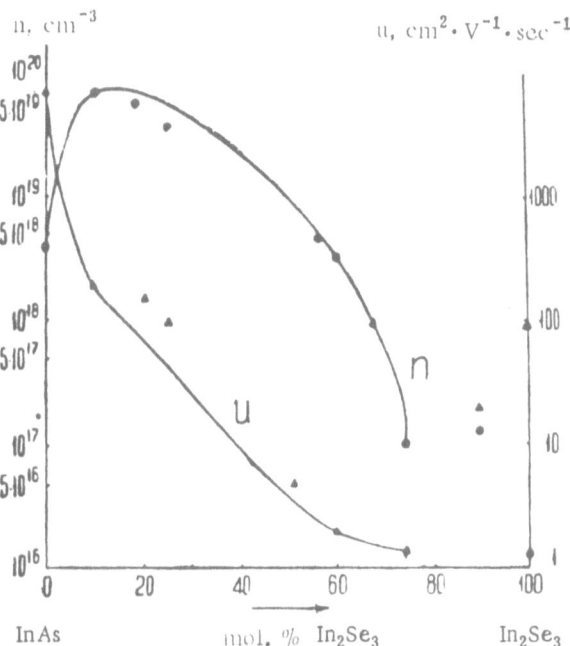

Fig. 9. Dependence of the carrier density and mobility on the composition of the InAs−In₂Se₃ system [10, 57].

Preliminary electrical measurements [55] have shown that the addition of In_2Se_3 to InP sharply raises the electrical conductivity by more than one order of magnitude but on further increase of the selenide concentration in the alloy the electrical conductivity gradually decreases. The temperature dependence of the electrical conductivity are similar to those obtained for $Ga_3As_3-Ga_2Se_3$ [9]. The compositions with high electrical conductivity are degenerate and the value of σ is almost independent of temperature (Fig. 7). The electrical conductivity of alloys with compositions x < 0.5 is lower but it increases with increase of temperature.

A preliminary study of the $(InP)_x-(InSe)_{1-x}$ system has shown that alloys of compositions x = 0.9, 0.7, and 0.5 have the zinc blende type structure with the same lattice parameter of 5.86_8 A.

b) Systems Based on Indium Arsenide. In all three systems based on indium arsenide wide miscibility (solid-solution) regions have been detected. We shall consider each of these systems separately.

$In_3As_3-In_2S_3$ System. There has been little work on this system. Results are available [52] indicating the possibility of solid solution formation near the indium arsenide composition in the range x = 1-0.5. The zinc blende structure is retained, and the lattice parameter decreases from 6.05_6 A to 6.02_5 A.

$In_3As_3-In_2Se_3$ System. This system has been studied in fairly great detail by investigators in various countries. In 1958, the present authors [12] used x-ray diffraction analysis to establish the solubility in the system, immediately after preparation, in the range 81.8 equimol. % (up to 40 mol. % InAs). Prolonged annealing and homogenization by annealing under pressure [56, 22] can extend this solubility region to 90 equimol. %. These results were confirmed in 1960 by Hahn and Thiele [58] and by Woolley and Keating in 1961 [59]. Radautsan [56] used thermal analysis to investigate the phase diagram of this system (Fig. 8).

Investigations of the electrical, galvanomagnetic [57, 59] and optical [68, 59] properties of alloys of this system have shown that all the alloys are semiconductors with the forbidden band width within the range of values found for the initial compounds. According to Woolley and Keating [59], the forbidden band width variation with composition is not continuous but is represented by a curve with a kink.

The addition of In_2Se_3 to InAs increases the carrier density, markedly pushes up the electrical conductivity and considerably reduces the mobility (Fig. 9).

The thermal conductivity of alloys in this system decreases from indium arsenide, first very rapidly (Fig. 10) and then more slowly, in the direction of In_2Se_3 [69]. Since the thermal conductivity of all the compositions increases on cooling, we may assume that heat is transported mainly by lattice vibrations. The linear

13

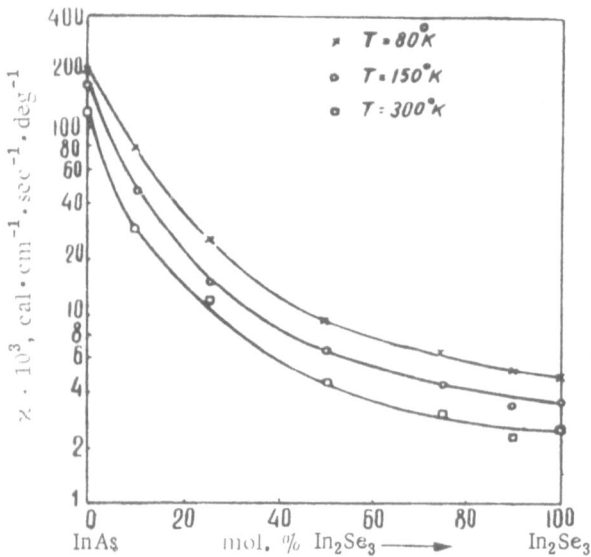

Fig. 10. Thermal conductivity of alloys in the
InAs—In$_2$Se$_3$ system [69].

expansion coefficient increases from indium arsenide to indium selenide. Hahn and Thiele [58] also established solid solutions in the InAs-InSe system in the concentration range 57 mol. %. The existence of the compound In$_2$AsSe with a lattice parameter 6.02 A is postulated. The low pycnometric density indicates a highly defective state of these alloys.

In$_3$As$_3$—In$_2$Te$_3$ System. The continuous solubility of this system was discovered first in 1958 by the present authors [60] and several months later it was confirmed by Woolley and Smith [41]. Two years later Hahn and Thiele obtained similar results. Figure 11 shows the variation of the lattice parameter with composition according to results of several authors [10, 41, 58]. As in similar systems, there is some deviation from Vegard's law near InAs. All compositions have the sphalerite structure with the lattice parameters varying from 6.05$_6$ A to 6.16 A.

The alloys with x = 0.57-0.5 exhibit a certain broadening of the Debye pattern lines, which may indicate some inhomogeneity of these compositions due to microliquation.

Thermal analysis has made it possible to plot the phase diagram for alloys of this system [10]. Clear critical points, characteristic of solid solutions, have been obtained. Compositions close to indium arsenide have additional weak effects at temperatures of 600-650°C, which may be due to either insufficient homogenization of these alloys or due to polymorphic transitions at these temperatures. A suggestion was made by Woolley and Pamplin [61] that ordering may occur in this system. Pamplin [70] compared nondefect and defect alloys exhibiting the sphalerite structure (defects in the lattice are denoted by □) and concluded that ordering is most likely at the compositions:

1. In$_3$□ AsTe$_3$ (x = 0.25),

2. In$_5$□ As$_3$Te$_3$ (x = 0.50),

3. In$_7$□ As$_5$Te$_3$ (x = 0.625).

There are no experimental data of the existence of ordering in alloys of this system.

Investigations of the electrical and galvanomagnetic properties [10] showed that in this system, as in the one discussed just before it, the electrical conductivity σ and the carrier density n increase strongly on the addition of small amounts of indium telluride to indium arsenide. Further increase of the In$_2$Te$_3$ concentration produces a gradual decrease of σ and n.

The carrier mobility decreases rapidly when the indium arsenide concentration is reduced in the alloy (on the reduction of x), varying by four orders of magnitude (Fig. 12).

Measurements of the temperature dependence of the thermoelectric power [61, 63] showed that at all compositions there is a monotonic increase with temperature.

The thermal conductivity of alloys of this system [62] has a minimum value at x = 0.5 (Fig. 12).

Optical investigations have made it possible to obtain some information on the variation of the forbidden band width. Alloys of three compositions were investigated [68] and the values of ΔE, calculated from absorption edge, were found to lie between the corresponding values of the initial binary compounds. A later study, carried out by the diffuse reflection method [61], also yielded data on the variation of the forbidden band in the form of a curve with a minimum near InAs.

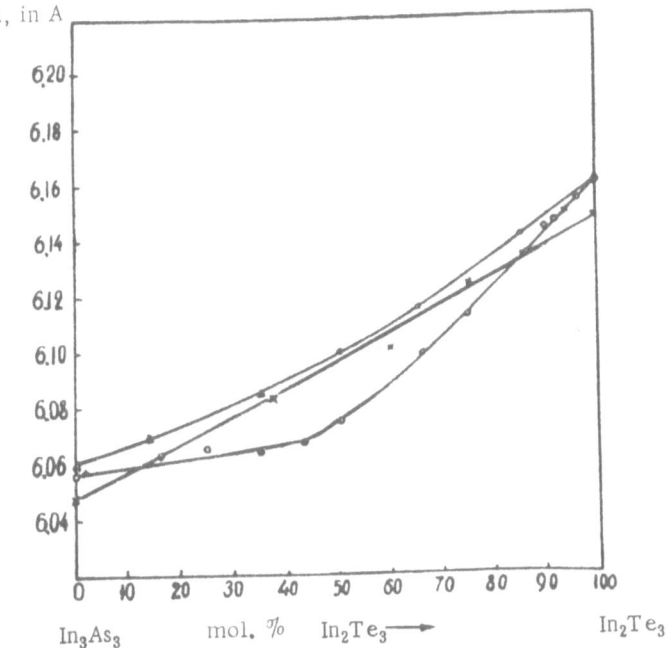

Fig. 11. Variation of the lattice parameter of the $In_3As_3-In_2Te_3$ system with its composition [10, 41, 58].

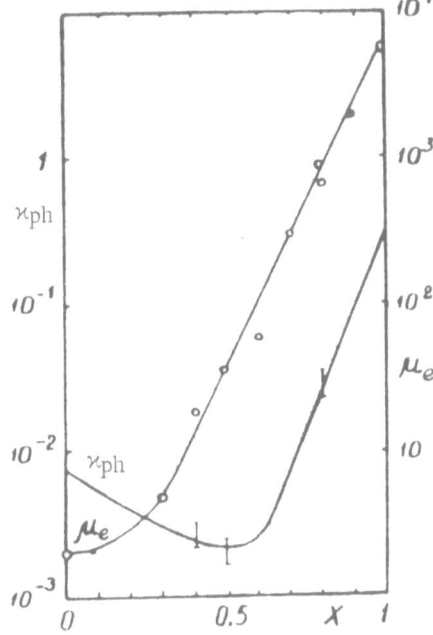

Fig. 12. Variation of the carrier mobility (μ_e) and of the thermal conductivity $\varkappa_{ph}$ of the $(InAs)_{3x}-(In_2Te_3)_{1-x}$ system with its composition.

Hahn and Thiele [58] detected $(InAs)_x$ $-(InTe)_{1-x}$ solid solutions with the sphalerite structure in the concentration region $x = 1-0.5$.

c) Systems Based on Indium Antimonide. Preliminary investigations of the systems $InSb-In_2B_3^{VI}$ [52], where B^{VI} is sulfur, selenium, or tellurium, showed that these systems do not form solid solutions in a wide range of concentrations and therefore the solubility is possible only in narrow ranges of concentrations near the intial compounds.

A more detailed study of the system $In_3Sb_3-In_2Se_3$ [59] showed that near indium antimonide sphalerite-type solid solutions are formed in the range 2.5 equimol. %.

In the $InSb-In_2Te_3$ system [64] the solubility range is somewhat wider: ≈ 15 equimol. %. The cited paper quotes also some electrical and optical properties of alloys of this system. In other work [65, 71, 72], it was shown that a new chemical compound, with the rock-salt structure and lattice parameter of 6.12_8 A, is formed along the tie-line $(InSb)_x-(InTe)_{1-x}$ at the composition $x = 0.25$ (In_4SbTe_3).

A similar compound is not formed in the $InSb-InSe$ system but the replacement of tellurium with selenium in the compound In_4SbTe_3 has produced solid solutions along the tie-line $In_4SbTe_{3x}Se_{3(1-x)}$ in the range $x = 1-0.75$ [73]. The lattice parameter then varies from 6.12_8 A to 6.06_5 A.

A recent study of the system $(InSb)_x$ $-(InTe)_{1-x}$ near indium antimonide showed that also along this tie-line solid solutions with the zinc blende type structure are formed at $x = 1-0.85$. Some structural, thermal, and electrical properties of these solid solutions were investigated [72]. The same investigation showed that it is possible to form solid solutions near indium antimonide along other tie-lines of the indium–antimony–tellurium ternary system.

There is also some solubility of indium selenide (InSe) in indium antimonide [73].

D. A^{II}B^{VI}–A$_2^{III}$B$_3^{VI}$ SYSTEM

Hahn and Frank [75] investigated ternary tetrahedral phases and detected a new group of substances $A^{II}B_2^{III}C_4^{VI}$, which they studied by x-ray diffraction. These substances were made from binary compounds $A^{II}B^{VI}$

15

TABLE 7. Some Properties of Tetrahedral Phases of the $A^{II}B_2^{III}C_4^{VI}$ Type

No.	Substance	Color	Lattice parameters in kXU			Density (exper.), g/cm^3	Density (x-ray). g/cm^3	Space group or structure type	
			a	c	c/a				
1	$ZnAl_2S_4$	light gray*	3.75_6	6.13	1.63_2	2.65	2.72_7	W	
2	$CdAl_2S_4$	black	5.55_3	10.30	1.82	3.04	3.06_2	S_4^2	
3	$HgAl_2S_4$	"	5.47_7	10.24	1.87_3	4.08	4.11_2	S_4^2	
4	$ZnAl_2Se_4$	"	5.49_2	10.88	1.98_2	4.37	4.37_6	S_4^2	
5	$CdAl_2Se_4$	"	5.73_5	10.66	1.85_3	4.50	4.54_2	S_4^2	
6	$HgAl_2Se_4$	"	5.69_6	10.7_2	1.88_2	5.0_2	5.05_3	S_4^2	
7	$ZnAl_2Te_4$	"	5.09_4	12.0_3	2.03_3	4.9_1	4.95_5	S_4^2	D_{2d}^{11}
			5.09_4	6.01_5	1.01_9			D_{2d}^1	D_{2d}^9
8	$CdAl_2Te_4$	"	5.99_9	12.1_9	2.03	5.1_0	5.09_9	S_4^2	D_{2d}^{11}
			5.99_9	6.09_5	1.01_5			D_{2d}^1	D_{21}^9
9	$HgAl_2Te_4$	"	5.99_2	12.09	2.01_7	5.7_9	5.81_6	S_4^2	D_{2d}^{11}
			5.99_2	6.04_5	1.00_9			D_d^1	D_{2d}^9
10	$ZnGa_2S_4$	white	5.26_3	10.4_2	1.97_9	3.75	3.80_3	S_4^2	D_{2d}^{11}
11	$CdGa_2S_4$		$5,56_6$	10.0_6	1.80_8	3.9_7	4.03_2	S_4^2	
12	$HgGa_2S_4$		$5,49_6$	10.2_1	1.86	4.95	5.00_2	S_4^2	
13	$ZnGa_2Se_4$		$5,48_5$	10.9_7	2.00	5.1_3	5.21_5	S_4^2	D_{2d}^{11}
14	$CdGa_2Se_4$		$5,73_1$	10.7_1	1.87_0	6.28	5.32_6	S_4^2	
15	$HgGa_2Se_4$	black	$5,70_3$	10.76	1.88_6	6.10	6.18_5	S_4^2	
16	$ZnGa_2Te_4$	"	$5,92_5$	11.8_5	2.00	5.5_7	5.67_4	S_4^2	D_{2d}^{11}
17	$CdGa_2Te_4$	"	$6,08_1$	11.7_9	1.93	5.6_3	5.77_1	S_4^2	
18	$HgGa_2Te_4$	"	$6,00_5$	12.0_1	2.00	6.42	6.48_1	S_4^2	
19	$ZnIn_2Se_4$	"	$5,69_0$	11.40	2.00	5.36	5.44_3	S_4^2	
20	$CdIn_2Se_4$	"	$5,80_5$	5.80_5	1.00	5.5_4	5.54_8	D_{2d}^1	
21	$HgIn_2Se_4$	"	$5,75_2$	11.7_3	2.04_7	6.26_2	6.33_1	S_4^2	
22	$ZnIn_2Te_4$	"	$6,11$	12.2_2	2.00	5.8_2	5.82_6	S_4^2	
23	$CdIn_2Te_4$	"	$6,19_2$	12.3_3	2.00	5.8_8	5.92_4	S_4^2	
24	$HgIn_2Te_4$	"	$6,17_4$	12.3_5	2.00	6.3_4	5.59_5	S_4^2	

*In this compound, the tetrahedral wurtzite structure exists only above 1000°C.

TABLE 8. Interaction in the $A_3^{II}B_3^{VI} - A_3^{III}B_3^{VI}$ Systems

System	Solid solution range	References
$Ga_2Se_3 - Zn_3Se_3$	Continuous solid solutions	3
$Ga_2Te_3 - Zn_3Te_3$	Solid solutions with gap at 55-73 equimol. % Ga_2Te_3	4,23
$Ga_2Te_3 - Cd_3Te_3$	Solid solutions with gap at 42-73 and 83-87 equimol. % Ga_2Te_3	23
$Ga_2Te_3 - Hg_3Te_3$	Solid solutions with gap at 42-73 equimol. % Ga_2Te_3	23
$In_2Se_3 - Cd_3Se_3$	Homogeneous region with ZnS structure in the 25.7 equimol. % range	5,76
$In_2Se_3 - Hg_3Se_3$	Solid solutions in the 75 equimol. % range on the Hg_3Se_3 side	77
$In_2Te_3 - Zn_3Te_3$	Continuous solid solutions with ordering in the 60-90 equimol. % In_2Te_3 range	78
$In_2Te_3 - Cd_3Te_3$	Solid solutions with gap at 50-72 equimol. % and ordering at 72-85 equimol. % In_2Te_3	78
$In_2Te_3 - Hg_3Te_3$	Solid solutions with gap at 48-60 equimol. % and ordering at 25-45 and 60-90 equimol. % In_2Te_3	78

Note: in Table 8, the compounds $A^{II}B^{VI}$ like the compounds $A^{III}B^{V}$ in Table 5, are referred to as $A_3^{II}B_3^{VI}$ to retain the equimolecular relationship.

and $A_2^{III}B_3^{VI}$, mixed in stoichiometric proportions, compacted into tablets, and then heated for 12-24 hours at a high temperature (700-1100°C) [11]. Homogenization of indium and gallium compounds was carried out at 600°C for 12 hours and that of mercury compounds took several weeks.

Table 7 lists the formulas and the x-ray diffraction results for substances of this type with tetrahedral structure.

The ternary oxides ($ZnAl_2O_4$, $CdGa_2O_4$, etc.) and indium-based sulfides ($ZnIn_2S_4$, $CdIn_2S_4$, and $HgIn_2S_4$) have spinel structures formed because the octahedral coordination is preferred in the corresponding binary compounds (oxides and sulfides) [11]. In the opinion of the investigators, all these tetrahedral phases may be regarded as superstructures in the solid-solution region of binary compounds $A^{II}B^{VI}$ and $A_2^{III}B_3^{VI}$.

In other work, the interaction of these compounds was considered from the point of view of their isomorphism [4, 5].

In some $A_3^{II}B_3^{VI} - A_2^{III}B_3^{VI}$ systems, the solubility regions were investigated in greater detail [23, 76, 77, 78]. Table 8 lists the results of these investigations.

Continuous solid solutions with the sphalerite structure were found in the $Ga_2Se_3 - Zn_3Se_3$ system [4]. The investigators have remarked that all compositions exhibit some deviation from the linearity in the dependence of the lattice parameter on composition; the deviation is in the direction of the larger parameter (that of zinc selenide).

Woolley and Ray [23] investigated $Ga_2Te_3 - A_3Te_3$ alloys, where A is zinc, cadmium, or mercury. In all three systems, they found wide solubility regions near the initial compounds, with discontinuities in the middle range of concentrations. The width of the two-phase region is from 18 to 35 equimol. %. Ordering is observed (with formation of the chalcopyrite structure) in alloys containing 75 equimol. % Ga_2Te_3. This result is in agreement with the data obtained by Hahn et al. [11] (cf. Table 7, alloys 16, 17, and 18). Figure 13 shows a plot of the variation of the lattice parameter with composition for the system cadmium telluride—gallium telluride. Two miscibility gaps, at 42-73 and 83-87 equimol. % Ga_2Te_3, are observed, as well as a region of ordering at 73-83 equimol. % Ga_2Te_3.

The forbidden band width of the system $Zn_3Te_3 - Ga_2Te_3$, found from measurements of the optical properties, varies continuously between 2.18 eV and 1.2 eV. Bearing in mind the difficulties of homogenization, we can say that in this system solid solutions exist at all concentrations.

TABLE 9. Some Properties of Semiconducting $A^{II}B_2^{III}C_4^{VI}$ Phases

Substance	Formation temperature, °C	Resistivity at 25°C ohm-cm	Cond. type	ΔE, eV (elec. meas.)	ΔE, eV (opt. meas.)	References
$ZnIn_2Se_4$	980 ± 10	—	—	2.6	—	79,80
$CdIn_2Se_4$	915 ± 5	3.10^{-3}	n	1.45	1.3	79,76
$HgIn_2Se_4$	830 ± 10	—	—	0.6	—	79,80
$ZnIn_2Te_4$	800 ± 5	3.10^7	—	1.14—1.4	0.86—1.15	79,80
$CdIn_2Te_4$	785 ± 2	60	n	0.9	0.92—1.08	79,80
$HgIn_2Te_4$	708 ± 2	600	n	1.25	0.86	79,80,82

Goryunova et al. [5] found wide regions of homogeneity in the system $CdSe-In_2Se_3$. Alloys of this system were investigated in greater detail by Kolomiets and Mal'kova [76]. Measurements of the electrical properties of $CdIn_2Se_4$ showed that this substance has semiconducting properties and that its forbidden band width is 1.45 eV. Introduction of impurities affects strongly the electrical conductivity of this material.

Radautsan and Gavrilitsa [77] reported a study of the system mercury selenide—indium selenide. Sphalerite solid solutions were detected in the range 75 equimol. % on the mercury selenide side, the lattice parameter varying within the limits 6.07-5.86 A. No ordering of alloys in this system was found. The disagreement with the data of Hahn et al. [11] may be due to the difference in the methods of preparation and possibly insufficient homogeneity.

Woolley and Ray [78] reported the existence of wide solubility ranges in the systems $In_2Te_3-A_3Te_3$, where A is zinc, cadmium, or mercury. As in the case of systems based on gallium telluride [23], ordering regions with the chalcopyrite structure are observed in the middle range of concentrations while in alloys based on CdTe and HgTe two-phase regions are found at 22 and 12 equimol. %, respectively (Table 8). X-ray diffraction studies agree quite well with the results of Hahn and Frank [75] (Table 7, alloys 22, 23, and 24). Optical measurements have been used to deduce the monotonic variation of the forbidden band width in the systems $Zn_3Te_3-In_2Te_3$ (2.18-1.1 eV) and $Cd_3Te_3-In_2Te_3$ (1.48-1.1 eV). Mason and O'Kane [79] carried out thermal analysis of $A^{II}In_2B_4^{VI}$ alloys, where A is zinc, cadmium, or mercury, and B is selenium or tellurium. The only material that melts congruently is $HgIn_2Te_4$, the others are formed by a peritectic reaction. Table 9 lists some properties of these substances.

Mason and O'Kane [79] list the electrical properties of $CdIn_2Te_4$ prepared by the zone leveling method. At room temperature, n-type samples were found to have a carrier density of 10^{14} cm^{-3} and an electron mobility of 4000 cm$^2\cdot$V$^{-1}\cdot$sec^{-1}.

Recent investigations [83] have shown that solid solutions can be formed between complex phases $A^{II}B_2^{III}C_4^{VI}$. Thus along the tie-line $CdIn_2Se_4-CdIn_2Te_4$ wide solubility regions have been detected with the lattice parameter varying continuously with the composition.

Beun, Nitsche, and Lichtensteiger [21] studied the photoelectric properties of a series of ternary chalcogenides of $A^{III}B_2^{III}C_4^{VI}$ type, prepared by the transport reaction method. Table 10 lists the data on the external form and some semiconducting properties of these materials. These investigators point out the promising photosensitivity of such substances as $ZnIn_2S_4$ and $CdGa_2Se_4$.

E. SYSTEMS BASED ON $A^{II}B^{IV}C_3^{VI}$ PHASES

These substances are the ternary analogs of defect compounds of the $A_2^{III}B_3^{VI}$ type. Some compounds of the $A^{II}B^{IV}C_3^{VI}$ type may not exist, but in conjuction with the binary compounds $A^{II}C^{VI}$ and $A_2^{III}B_3^{VI}$ we may expect solid solutions to form over a certain range of concentrations.

TABLE 10. Some Properties of Substances of the $A^{II}B_2^{III}C_4^{VI}$ Type, Prepared by the Transport Reaction Method

Substance	External form	Dimensions, in mm	ΔE, eV, from photoelec. meas.	Maximum sensitivity at λ_{max} (μ)	Resistivity, ohm·cm		Electron mobility μ, cm^2·V^{-1}·sec^{-1}
					dark	illum.	
$ZnIn_2S_4$	yellow thin plates	10×10×0.1	2.6	0.48	$1 \cdot 10^{14}$	$2 \cdot 10^8$	19
$CdIn_2S_4$	red octahedra	5×5×5	2.3	0.54	$1 \cdot 10^8$	$3 \cdot 10^4$	56
$HgIn_2S_4$	black octahedra	1×1×1	2.0	0.62	$2.7 \cdot 10^6$	$1.8 \cdot 10^6$	1.7
$ZnIn_2Se_4$	black prisms	5×1×1	1.82	0.58	$2.4 \cdot 10^7$	$1.8 \cdot 10^4$	35
$CdIn_2Se_4$	black prisms	5×1×1	1.72	0.77	$8 \cdot 10^5$	$7 \cdot 10^5$	22
$ZnGa_2S_4$	white polyhedra	0.1×0.1×0.1	—	0.39	$4 \cdot 10^{11}$	$1.5 \cdot 10^{10}$	—
$CdGa_2S_4$	colorless prisms	6×2×2	3.44	0.35	$8 \cdot 10^{13}$	$2.5 \cdot 10^8$	60
$HgGa_2S_4$	yellow needles	8×0.3×0.3	2.84	0.49	$1 \cdot 10^{10}$	$7 \cdot 10^4$	—
$ZnGa_2Se_4$	orange polyhedra	1×0.5×0.5	—	0.57	$2.5 \cdot 10^{12}$	$1 \cdot 10^3$	—
$CdGa_2Se_4$	red prisms	4×1×1	2.43	0.48	$4 \cdot 10^{11}$	$1.1 \cdot 10^5$	33
$HgGa_2Se_4$	black needles	10×0.5×0.5	1.95	0.62	$1.4 \cdot 10^7$	$2.7 \cdot 10^4$	600

Radautsan and Ivanova [84] reported the results of studies of the following systems:

1) $ZnGeSe_3 - ZnSe$;

2) $ZnGeSe_3 - Ga_2Se_3$;

3) $ZnGeSe_3 - In_2Se_3$.

X-ray diffraction and microstructure analyses have shown that in all three systems there are wide homogeneous regions within which alloys have the zinc blende structure. In the first system, the solid solution region covers the whole range of concentrations, and the lattice parameter varies but little. In the second and third systems this region is not less than 50 mol. %, with the lattice parameter varying from 5.64_5 A ($ZnGeSe_3$) to 5.46_3 A (50 mol. % Ga_2Se_3), or 5.69_8 A (50 mol. % In_2Se_3), respectively. We must note the following interesting phenomenon observed in studies of alloys of $ZnGeSe_3 + Ga_2Se_3$ composition (x-ray diffraction studies were carried out by M. M. Markus, a member of the staff of the Institute of Physics and Mathematics of the Moldavian Academy of Sciences). The Debye diffraction fractions of this substance show that after annealing, the lines with even indices are accompanied by lines of equal intensity but shifted toward larger angles by 0.02 [on the $(\sin \theta)/\lambda$ scale]. In alloys with high concentrations of Ga_2Se_3, as in the compound Ga_2Se_3 itself [85], broadening and shift of the lines with odd indices has been observed.

In 1959 Hahn reported [86] compounds of the $A_2^{II}B^{IV}C_4^{VI}$ type which he considered as alloys along the sections $A^{II}C^{VI} - B^{IV}C_2^{VI}$. The results of studies of the structure and determinations of the density of some of these substances are listed in Table 11.

The substances Zn_2GeSe_4 crystals with the sphalerite structure have a wide region of homogeneity along the sections $ZnS - GeS_2$ and $ZnSe - GeSe_2$, respectively. The results of an investigation of $ZnGeSe_4$ are in good agreement with the data of Radautsan and Ivanova [84].

It is suggested that all alloys listed in Table 11 are semiconductors.

TABLE 11. Some Properties of $A_2^{II}B^{IV}C_4^{VI}$– Type Phases

Substance	Structure	Lattice parameters, in A			Density, g/cm^3	
		a	c	c/a	calc.	exper.
Zn_2GeS_4	sphalerite	5.43_6	—	—	3.42_7	3.26_0
Zn_2GeSe_4	"	5.64_6	—	—	4.78_9	4.53_2
Cd_2GeS_4	hexagonal-rhombohedral	7.1_3	35.1	4.9_2	4.11_5	3.82_4
Cd_2GeSe_4	"	7.4_1	36.2	4.8_9	5.44_4	5.19_4
Hg_2GeS_4	"	7.1_7	34.9	4.68_6	5.78_9	6.61_1
Hg_2GeSe_4	thiogallate	5.69_1	11.28_0	1.98_2	7.17_9	7.09_0

TABLE 12. Some Properties of $A_2^{I}B^{II}C_4^{VI}$– Type Compounds

Substance	Modification	Color	Lattice period, in A	Density, g/cm^3
Cu_2HgI_4	α	brown	6.103 ± 0.005	6.009
	β	red	$a = 6.08 \pm 0.005$ $c = 12.218 \pm 0.005$	
Ag_2HgI_4	α	red	6.383 ± 0.005	5.930
	β	yellow	$a = 6.304 \pm 0.005$ $c = 12.608 \pm 0.005$	

F. TERNARY COMPOUNDS OF $A_2^{I}B^{II}C_4^{VI}$ TYPE

Representative defect phases of this type are the compounds Cu_2HgI_4 and Ag_2HgI_4. The high-temperature α modifications of these compounds crystallize with structures similar to sphalerite, but with one-quarter of the randomly distributed vacancies in the cation sublattice [87-89].

The low-temperature β modifications differ somewhat although both compounds crystallize with tetrahedral structures. The difference is in the positions of the metal atoms in the lattice: in β-Cu_2HgI_4 they are in the form of layers, while in β-Ag_2HgI_4 they are in tetrahedral spaces empty spaces in the cubic close packing of iodine atoms.

The temperature of the $\alpha \rightarrow \beta$ transition for the copper compound is 70°C and for the silver compound it is 50°C.

Table 12 lists some physico-chemical properties of these substances.

The electrical properties of these compounds have not been investigated. Both compounds form solid solutions with each other, which crystallize in the β-Cu_2HgI_4 structure with a random distribution of copper and silver atoms in copper atom positions [88, 90].

DISCUSSION OF THE RESULTS

An examination of the experimental results on diamond-like defect semiconductors indicates that they have properties with many features in common. This is found in the series of analogs of binary (Table 3), ternary (Tables 7, 9, 10, 11) and, very likely, more complex compounds.

The relationship between diamond-like defect semiconductors and nondefect tetrahedral phases is obvious. The similarities and differences between these two groups of substances appear, for example, on examination of the mixed isoelectronic series (Table 1). The specific properties of defect semiconductors may be found by

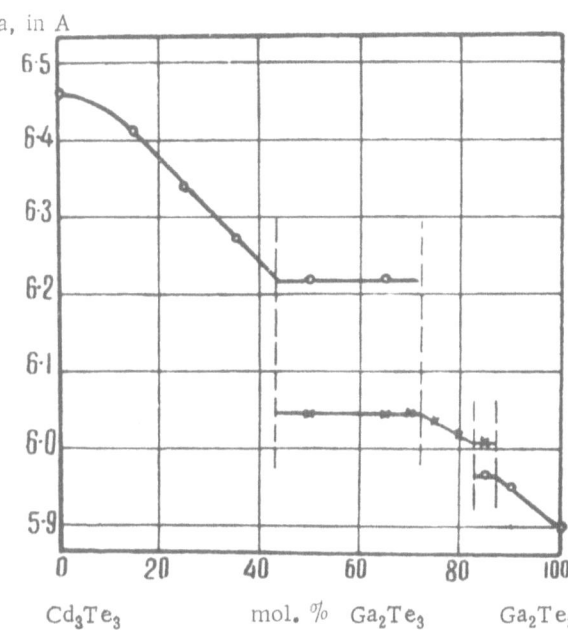

Fig. 13. Variation of the lattice parameter of the Cd_3Te_3–Ga_2Te_3 system with its composition [23].

examination of the single-type isoelectronic series, including more complex phases such as Ga_2Se_3, $ZnGeSe_3$, $ZnGa_2GeSe_6$, in which the number of defects remains constant on increase of the total number of atoms. Further information may be obtained by comparing series of substances with the same number of defects and the same total number of atoms but belonging to different groups in the periodic system, for example Zn_2GeSe_4 and $ZnGa_2Se_4$, as well as by comparing substances in series having different numbers of defects, for example Ga_2Se_3, $ZnGa_2Se_4$, $ZnSe$.

Diamond-like defect substances easily form substitutional solid solutions, both of the isovalent and heterovalent type, when they interact between one another or when they interact with nondefect compounds with the ZnS structure. This makes it possible to obtain a wide range of substances from diamond-like nondefect semiconductors to those with maximum number of defects, including all the intermediate values of the number of defects and of the electron density. Thus, the data on the formation of a family of tetrahedral phases [1] in that part which includes defect substances are extended and made more precise.

The change in the number of defects and the resultant change in the electron density strongly affect the properties of the substances. The number of known single-type groups or systems of defect semiconductors is still small. However, general relationships in the variation of the principal properties (Figs. 5, 7, 9, 10, 12) have been found in the systems investigated, for example the $A^{III}B^V$–$A_2^{III}B_3^{VI}$ systems (GaAs–Ga_2Se_3, InP–In_2Se_3, InAs–In_2Se_3, InAs–In_2Te_3, and others).

The large number of "intrinsic" defects (cation vacancies), reaching 5.5×10^{21} cm^{-3} in $A_2^{III}B_3^{VI}$ compounds, disturbs the periodicity of the lattice and strongly distorts the potential field and therefore alters the physical properties. By varying the composition of these solid solutions, we can control the number of "intrinsic" defects, i.e., continuously vary the number of vacant sites in the zinc blende lattice. An exhaustive analysis of the electrical, galvanomagnetic, thermoelectric, and thermal properties of these alloys has yielded information on the influence of the "intrinsic" defects on the nature of carrier and phonon scattering.

The interaction of diamond-like defect substances has certain characteristic features. Nondefect compounds, for example $A^{III}B^V$, interact with one another to form substitutional solid solutions usually along the $A^{III}B^V$–$A_1^{III}B^V$ or $A^{III}B^V$–$A^{III}B_1^V$ tie-lines of the ternary systems A^{III}–A_1^{III}–B^V or A^{III}–B^V–B_1^V (cation or anion substitution). There are no homogeneous phases in other regions of such ternary systems.

The interaction of tetrahedral phases having "intrinsic" defects is different. Thus, for example, a wide homogeneous region is observed on the indium antimonide side in the indium–antimony–tellurium system [74]. Homogeneous phases are formed not only along the tie-lines InSb–In_2Te_3, InSb–InTe, but also in the intermediate regions between these tie-lines. In spite of different chemical composition, the alloys in these regions have the same structure and similar lattice parameters. By analogy with the explanation of the formation of the compound In_4SbTe_3 [60], we may assume that in these homogeneous phases both the monovalent and the trivalent indium is present. It is worth noting that a study of some of the electrical properties of alloys along the InSb–In_2Te_3 [64] and InSb–InTe [74] tie-lines indicated their common nature.

Some published work [58, 59, 73] indicates that wide homogeneous regions occur frequently in other similar systems, as shown by studies of some of the alloys along particular tie-lines.

The presence of solid solutions along the tie-lines $InAs-In_2A_3$ and $InAs-InA$, where A is selenium or tellurium, in the ternary systems indium–arsenic–selenium and indium–arsenic–tellurium indicates the presence of homogeneous regions based on indium arsenide.

A similar result was obtained for the indium–phosphorus–selenium system on investigation of the tie-lines $InP-In_2Se_3$ and $InP-InSe$.

The ordering, which is characteristic of many defect alloys, of the binary compounds Ga_2Se_3 and In_2Te_3 was studied exhaustively [34, 85].

In more complex defect alloys, the ordering appears normally on cation substitution and is very rare in anion substitution. Thus, in systems $A^{III}B^V-A_2^{III}B_3^{VI}$ the ordering was observed only in $GaP-Ga_2Se_3$. Recently, Woolley and Keating [91] reported possible ordering in the system $InAs-In_2Se_3$.

In the $A_2^{III}B_3^{VI}-A^{II}B^{VI}$ systems, the ordering has been observed for the majority of the alloys (Table 7, Fig. 13). Recent investigation indicate that in more complex quaternary systems the ordering is observed in defect substances. This appears particularly strongly in alloys based on the binary ordered compounds Ga_2Se_3 and In_2Te_3. The Debye powder patterns of the alloys in the solid solution region show a regular displacement of the lines, representing ordering, together with the main lines [83].

Especially worth mentioning are the observations, referred to in the above review of the experimental results, that the lines with even indices in the Debye patterns of annealed $ZnGeSe_3+Ga_2Se_3$ split into two. This result may be due to regular distribution of atoms of the quaternary system and of defect sites in the lattice. It would be interesting to obtain information on the influence of the ordering on the semiconducting properties of defect systems.

There are practically no data on the influence of impurities on the properties of defect materials and the preliminary results available so far cannot always be explained [34]. It is worth stressing that the formation of impurity and exciton levels is very difficult in defect compounds [34, 38].

In spite of the relative incompleteness of the data on complex defect alloys, we can suggest some possible practical applications. One of such applications may be in the manufacture of thermoelectric devices because some compositions may have the optimum relationship between the electrical conductivity, thermoelectric power, and thermal conductivity, which governs the thermoelectric efficiency. It is particularly important that some of these materials have high melting points and their semiconducting properties are nearly constant over a wide range of temperatures.

Solid solutions in which the absorption band edge varies strongly with the composition may be used to make optical filters for various parts of the spectrum. Some alloys have good photoelectric properties, which makes them promising as potential photoresistors and photocells.

A detailed study of the properites of solutions with high concentrations of the $A^{III}B^V$ compounds would be interesting in connection with their possible use as degenerate semiconductors for the manufacture of tunnel diodes.

The whole material cited in the present review and some of the general conclusions arrived at on diamond-like defect semiconductors far from exhaust the subject of these semiconductors.

Further studies of these substances should reveal new general relationships, establish the special properties of complex alloys, and extend the range of materials used in semiconductor technology.

LITERATURE CITED

1. N. A. Goryunova, Chemistry of Diamond-Like Semiconductors, Izd. LGU (1963).
2. H. Hahn and W. Klingler, Z. Anorg. Allgem. Chem. 259: 121 (1949); 260: 97 (1949).
3. J. A. A. Ketelaar, Z. Krist. 87: 436 (1934).
4. N. A. Goryunova, V. A. Kotovich, and V. A. Frank-Kamenetskii, Dokl. Akad. Nauk SSSR 103: 659 (1955).
5. N. A. Goryunova, V. A. Kotovich, and V. A. Frank-Kamenetskii, Zh. Tekh. Fiz. 25: 2419 (1955).
6. N. A. Goryunova, Collection: Problems in the Theory and Experimental Studies of Semiconductors and Semiconductor Metallurgy Processes, Izd. Akad. Nauk SSSR, Moscow (1955), p. 29.
7. J. Appel, Z. Naturforsch. 9a: 265 (1954).
8. N. A. Goryunova, V. S. Grigor'eva, B. M. Konovalenko, and S. M. Ryvkin, Zh. Tekhn. Fiz. 25: 1675 (1955).
9. D. N. Nasledov and I. A. Feltyn'sh, Fiz. Tverd. Tela 1: 565 (1959); Fiz. Tverd. Tela 2: 823 (1960).
10. S. I. Radautsan, Investigation of Some Semiconductor Solid Solutions Based on Indium Arsenide (author's abstract of a dissertation), Leningrad (1958).
11. H. Hahn, G. Frank, W. Klingler, A. D. Störger, G. Störger, Z. Anorg. Allgem. Chem. 279: 241 (1955).
12. N. A. Goryunova and S. I. Radautsan, Zh. Tekhn. Fiz. 28: 1917 (1958).
13. N. A. Goryunova and V. I. Sokolova, Izv. Moldavsk. Filiala Akad. Nauk SSSR, No. 3(69): 31 (1959).
14. S. I. Radautsan, I. A. Madan, I. P. Molodyan, and R. A. Ivanova, Izv. Moldavsk. Filiala Akad. Nauk SSSR, No. 3(69): 107 (1959).
15. N. A. Goryunova, Investigations of Semiconductor Chemistry (doctoral dissertation) (1958).
16. A. S. Borshchevskii and D. N. Tret'yakov, Fiz. Tverd. Tela 1: 1483 (1959).
17. S. D. Remenko, Izv. Akad. Nauk Moldav.SSR, No. 10(88): 76 (1961).
18. A. S. Borshchevskii, Synthesis of Semiconducting Materials and Physico-Chemical Investigations of Some of Them (author's abstract of dissertation for candidate's degree), Leningrad (1962).
19. G. Bush, P. Junod, E. Mooser, and H. Schade, Halbleiter und Phosphore (1958), p. 470.
20. D. R. Mason and J. S. Cook, J. Appl. Phys. 32: 475 (1961).
21. J. A. Beun, R. Nitsche, and M. Lichtensteiger, Physica 26: 647 (1960); 27: 448 (1961).
22. N. A. Goryunova, S. I. Radautsan, and V. I. Deryabina, Fiz. Tverd. Tela 1: 512 (1959).
23. J. C. Woolley and B. Ray, J. Phys. Chem. Solids 16: 102 (1960).
24. V. P. Zhuze, A. I. Zaslavskii, V. A. Petrusevich, V. M. Sergeeva, I. A. Smirnov, and A. I. Shelykh, Proc. Interatl. Conf. Semicond. Phys., held in Prague, 1960 (publ. Prague 1961), p. 871.
25. A. S. Borshchevskii, N. A. Goryunova, and N. K. Takhtareva, Zh. Tekhn. Fiz. 27: 1408 (1957).
26. G. V. Bokii, Introduction to Crystal Chemistry, Izd. MGU (1954).
27. S. A. Semiletov, Fiz. Tverd. Tela 3: 746 (1961).
28. A. I. Zaslavskii, V. M. Sergeeva, and I. A. Smirnov, Fiz. Tverd. Tela 2: 2884 (1960).
29. E. Kaver and A. Rabenau, Z. Naturforsch. 13a: 531 (1958).
30. M. S. Mirgalovskaya and E. V. Skudnova, Izv. Akad. Nauk SSSR, Otd. Tekhn. Nauk Met. i Toplivo, No. 4: 148 (1959).
31. W. Klemm and H. U. Vogel, Z. Anorg. Chem. 219: 45 (1934).
32. N. A. Goryunova, Zh. Vses. Khim. Obshchestva im. D. I. Mendeleeva 5: 522 (1960).
33. M. Hansen and K. Anderko, Structure of Binary Alloys, Gos. NTIzd. Lit. Chern. i Tsvetn. Metallurgii, Moscow (1962).
34. V. P. Zhuze, V. M. Sergeeva, and A. I. Shelykh, Fiz. Tverd. Tela 2: 2858 (1960).
35. I. Z. Fisher, Fiz. Tverd. Tela 1: 193 (1959).
36. G. Harbeke and G. Lautz, Z. Naturforsch. 13a: 771 (1958).
37. A. M. Andriesh and B. T. Kolomiets, this volume, p. 35.
38. E. F. Gross, B. V. Novikov, B. S. Razbirin, and L. S. Suslina, Opt. i Spektroskpiya 4: 569 (1959).
39. V. P. Mushinskii, Izv. Vuzov MVO SSSR, Fizika, No. 6: 130 (1960).
40. V. I. Gramatskii, Uch. Zap. Kazakhsk. Gos. Univ. 49: 119 (1961).
41. J. C. Woolley and B. A. Smith, Proc. Phys. Soc. (London) 72: 867 (1958).
42. V. S. Grigor'eva, Substitutional Solid Solutions in Some Semiconducting Compounds with the ZnS Structure (author's abstract of dissertation for candidate's degree) Leningrad (1960).

43. V. S. Grigor'eva, Zh. Tekhn. Fiz. 28:1670 (1958).

44. I. N. Ageeva, A. A. Vaipolin, and V. S. Grigor'eva, Abstracts of Papers Presented at the Fourth Conference on Crystal Chemistry, Moscow (1961), p. 15.

45. S. I. Radautsan and O. P. Derid, Izv. Moldovsk. Filiala Akad. Nauk SSSR, No. 3(69):105 (1960).

46. M. S. Mirgalovskaya and E. V. Skudnova, Zh. Neorgan. Khim. 4:113 (1959).

47. S. I. Radautsan and V. V. Negreskul, this volume, p. 104.

48. S. I. Radautsan, I. A. Madan, and R. A. Ivanova, Izv. Akad. Nauk Moldav.SSR, No. 10(88):98 (1961).

49. I. I. Kozhina, S. S. Tolkachev, A. S. Borshchevskii, and N. A. Goryunova, Vestn. Leningr. Univ. 4:122 (1962).

50. N. A. Goryunova and V. S. Grigor'eva, Zh. Tekhn. Fiz. 26:2157 (1956).

51. M. S. Mirgalovskaya and E. M. Komova, Collection: Problems of Metallurgy and Physics of Semiconductors, Izd. Akad. Nauk SSSR (1961), p. 138.

52. S. I. Radautsan, Czech. J. Phys. 12:382 (1962).

53. N. A. Goryunova and V. I. Sokolova, Izv. Moldovsk. Filiala Akad. Nauk SSSR No. 3(69):31 (1960).

54. S. I. Radautsan, I. A. Madan, I. P. Molodyan, and R. A. Ivanova, Izv. Moldavsk. Filiala Akad. Nauk SSSR, No. 3(69):107 (1960).

55. S. I. Radautsan and I. A. Madan, Izv. Akad. Nauk Moldav.SSR, No. 5, 95, 92 (1962).

56. S. I. Radautsan, Zh. Neorgam. Khim. 4:1121 (1959).

57. S. I. Raduatsan and B. E. Sh. Malkovich, Fiz. Tverd. Tela 3:3324 (1961).

58. H. Hahn and D. Thiele, Z. Anorg. Allgem. Chem. 303:147 (1960).

59. J. C. Woolley and P. N. Keating, Proc. Phys. Soc. (London) 78:1009 (1961).

60. N. A. Goryunova and S. I. Radautsan, Dokl. Akad. Nauk SSSR 21:847 (1958).

61. J. C. Woolley, B. R. Pamplin, and J. A. Evans, J. Phys. Chem. Solids 19:147 (1961).

62. D. B. Gasson, P. J. Holmes, I. C. Jennings, J. E. Parrott, and A. W. Penn, Proc. Internatl. Conf. Semicond. Phys., held in Prague, 1960 (publ. Prague 1961), p. 1032.

63. S. I. Radautsan and L. M. Manovets, Izv. Akad. Nauk Moldav.SSR, No. 10:88, 71 (1961).

64. J. C. Woolley, C. M. Gillett, and J. A. Evans, J. Phys. Chem. Solids 16:138 (1960).

65. N. A. Goryunova, S. I. Radautsan, and G. A. Kiosse, Fiz. Tverd. Tela 1:1858 (1959).

66. A. R. Regel', Collection: Problems of the Theory and Experimental Investigations of Semiconductors and Semiconductor Metallurgy Processes, Izd. Akad. Nauk SSSR (1955).

67. M. S. Mirgalovskaya and E. M. Komova, Problems of Metallurgy and Physics of Semiconductors, Izd. Akad. Nauk SSSR (1961), p. 138.

68. D. N. Nasledov, M. P. Pronina, and S. I. Radautsan, Fiz. Tverd. Tela 2:50 (1960).

69. L. I. Berger and S. I. Radautsan, Problems of Metallurgy and Physics of Semiconductors, Izd. Akad. Nauk SSSR (1961), p. 129.

70. B. R. Pamplin, Nature (London) 188:136 (1960).

71. G. A. Kiosse, T. I. Malinovskii, and S. I. Radautsan, Izv. Moldovsk. Filiala Akad. Nauk SSSR, No. 3(69):3 (1960).

72. I. P. Molodyan, S. I. Radautsan, and I. A. Madan, Izv. Akad. Nauk Moldav.SSR, No. 10(88):91 (1961).

73. S. I. Radautsan, V. V. Negreskul, and I. A. Madan, Akad. Nauk Moldav.SSR, No. 10(88):57 (1961).

74. I. P. Molodyan and S. I. Radautsan, this volume, p. 94.

75. H. Hahn and G. Frank, Z. Anorg. Allgem. Chem. 269:227 (1952).

76. B. T. Kolomiets and A. A. Mal'kova, Fiz. Tverd. Tela, Sbornik II:33 (1959).

77. S. I. Radautsan and E. I. Gavrilitsa, Izv. Akad. Nauk Moldav.SSR, No. 10(88):95 (1961).

78. J. C. Woolley and B. Ray, J. Phys. Chem. Solids 15:27 (1960).

79. D. R. Mason and D. F. O'Kane, Proc. Internatl. Conf. Semocond. Phys., held in Prague, 1960 (publ. Prague 1961), p. 1026.

80. G. Busch, E. Mooser, and W. B. Pearson, Helv. Phys. Acta 29:192 (1956).

81. D. F. Edwards and D. F. O'Kane, Bull. Am. Phys. Soc., Ser. II, 5:78 (1960).

82. G. Busch, P. Junod, E. Mooser, and H. Schade, Halbleiter und Phosphore, Braunschweig (1958), p. 470.

83. S. I. Radautsan, R. A. Maslyanko, and M. M. Markus, this volume, p. 101.

84. S. I. Radautsan and R. A. Ivanova, Akad. Nauk Moldav.SSR, No. 10(88):64 (1961).

85. A. A. Vaipolin and V. S. Grigor'eva, this volume, p. 49.

86. H. Hahn, Seventeenth International Congress, Munich (Short Abstracts), Munich (1959), p. 157.

87. R. Jaza and F. Hund, Z. Anorg. Chem. 579 : 13 (1948).

88. H. Hahn, G. Frank, and W. Klingler, Z. Anorg. Allgem. Chem. 279 : 271 (1955).

89. C. E. Olsen and P. M. Harris, Phys. Rev. 86 : 659 (1952).

90. L. Suchow and P. H. Keck, J. Am. Chem. Soc. 75 : 518 (1953).

91. J. C. Woolley and P. N. Keating, J. Less-Common Metals 3 : 194 (1961).

SOME TERNARY COMPOUNDS OF THE $A_2^I B^{IV} C_3^{VI}$ TYPE AND SOLID SOLUTIONS BASED ON THEM

G. K. Averkieva, A. A. Vaipolin, and N. A. Goryunova

The extension of the range of applications of semiconductors has made it necessary to search for new materials. In 1958-1960, the first papers appeared on the formation of a new class of ternary semiconducting compounds with tetrahedral binding [1, 6]. Attempts have been made to generalize the experimental results and to derive criteria of the semiconducting nature and formation of compex tetrahedral phases [1, 5, 7, 8].

One of the five types of "two-cation" ternary diamond-like systems is that denoted by $A_2^I B^{IV} C_3^{VI}$. The published data on these substances are collected together in Table 1.

The present work is a continuation of our studies of ternary compounds of the diamond group, and of isovalent and heterovalent substitution in these compounds.

Studies of the solid solutions of $A_2^I B^{IV} C_3^{VI}$ in $A^{III} B^V$ are of considerable interest because they provide an opportunity of varying the properties continuously.

Ternary compounds of the $A_2^I B^{IV} C_3^{VI}$ type were prepared, here A^I was copper or silver, B^{IV} germanium or tin, and C^{VI} sulfur, selenium, or tellurium.

We prepared all the ternary compounds of this type containing copper; of those containing silver, we were able to make only Ag_2SnS_3 and Ag_2SnSe_3. X-ray diffraction was used to investigate: (1) the deviation from stoichiometry in the compound Cu_2GeSe_3, (2) the structure of Cu_2GeS_3, and (3) some solid solutions obtained by isovalent or heterovalent substitution.

Synthesis was carried out in evacuated quartz ampoules by melting together elements in their stoichiometric ratios. Some samples were subjected to heat treatment.

Microstructure analysis was used to find approximately the phase composition of the samples. The microhardness was measured with a PMT-3 type instrument, usually with a 50 g load. Thermal analysis was employed to determine the melting points (the crystallization ranges). Debye powder patterns were obtained for all the investigated samples.

1. TERNARY COMPOUNDS

The results obtained for the compounds are listed in Table 2 (where A^I denotes copper).

Table 2 shows the dependence of the melting point, the lattice parameter, and of the microhardness on the position of the compound in Mendeleev's table (represented by the sum of atomic numbers). It shows that with increase of the sum of atomic numbers the strength of bonds in the crystal lattice decreases, the lattice parameter increases, the melting point is reduced, and the microhardness becomes lower.

Measurements of the electrical properties carried out soon after synthesis showed that the compound Cu_2GeSe_3 had p-type conduction with a carrier density $n = 2.2 \times 10^{17} cm^{-3}$ and a conductivity $\sigma = 0.43 \, ohm^{-1} \cdot cm^{-1}$.

TABLE 1

Formula	Structure type	Lattice parameter			Density (g/cm³)		Micro-hardness kg/mm²	Melting point, °C	References
		a [A]	c [A]	c/a	x-ray	exper.			
Cu₂GeS₃	ZnS-type	5.32	10.41	1.958	4.43	4.339	—	—	[2]
	ZnS	5.30	—	—	4.39	4.36	607	948	[9]
Cu₂GeSe₃	ZnS-type	5.59	10.968	1.96	5.626	5.464	—	—	[2]
	tetragonal	5.58	10.96	—	—	—	—	788	[6]
	ZnS	5.55	—	—	5.65	5.65	532	770	[9]
Cu₂GeTe₃	ZnS-type	5.916	11.853	2.004	6.168	6.049	—	—	[2]
	ZnS	5.95	—	—	6.11	6.14	473	504	[9]
Cu₂SnS₃	ZnS	5.43	—	—	4.74	4.74	550	845	[9]
Cu₂SnSe₃	ZnS	5.69	—	—	5.79	5.74	514	696	[9]
Cu₂SnTe₃	ZnS	5.70	—	—	—	—	—	709	[6]
		6.04	—	—	6.31	6.23	412	407	[9]

TABLE 2

| Formula | Structure type | Lattice parameter | | Microhardness of princ. phase kg/mm^2 | Melting point, °C |
		a, kXU	c, kXU		
Cu_2GeS_3	tetragonal	5.316	5.205	464 ± 68	955
Cu_2SnS_3	ZnS	5.425	—	283 ± 30	855
Cu_2GeSe_3	tetragonal	5.578	5.474	391 ± 35	788
Cu_2SnSe_3	ZnS	5.676	—	256 ± 20	697
Cu_2GeTe_3	tetragonal	5.947	5.923	295 ± 15	492
Cu_2SnTe_3	ZnS	6.036	—	201 ± 20	411

After zone recrystallization, the compound Cu_2GeSe_3 retained its tetragonal lattice without any great change in its lattice parameters.

The compounds based on silver did not have the ZnS-type lattice. We shall report on them in greater detail in the near future.

2. INVESTIGATION OF THE COMPOUND Cu_2GeSe_3 OF NONSTOICHIOMETRIC COMPOSITION

Goryunova [10] investigated the problem of the formation of homogeneous phases along the Cu_2GeSe_3—Ge tie-line. This investigation showed that germanium was soluble in the compound up to 14% and that this transformed the tetragonal lattice into a cubic one.

We investigated the deviations from stoichiometry in the compound Cu_2GeSe_3 along the tie-line Cu_2Se—$GeSe_2$ in the Cu—Ge—Se system (Fig. 1) using a large number of samples. The microstructure and microhardness of these samples were measured. The phase composition was analyzed by x-ray diffraction. Over the range of concentrations from alloy No. 2 to alloy No. 12 the lattice was tetragonal.

It was found that near alloy No. 2 the structure changed: a new phase with a very different microhardness and of low-symmetry structure appeared. On deviating from stoichiometry in either direction we found that sections of the samples showed the presence of a second phase of different microhardness, the amount of which increased with the deviation from stoichiometry. This was confirmed by Debye powder patterns.

In recording x-ray diffraction patterns of fired samples with an excess of Cu_2Se, a discontinuous change of the lattice parameters, to a = 5.563 kXU and c/2 = 5.484 kXU, was observed (Table 2).

In the investigation of the fired samples with a deviation from stoichiometry in the direction of $GeSe_2$, a new phase appeared with the parameters a = 5.620 kXU and c/2 = 5.466 kXU and the amount of this phase increased with increase of the deviation.

The fired samples were annealed. X-ray diffraction phase analysis of the annealed samples indicated the presence of a principal phase with the parameters corresponding to Cu_2GeSe_3 (Table 2), in addition to lines whose intensity increased with the deviation from stoichiometry.

We concluded, therefore, that in the compound Cu_2GeSe_3 the region of homogeneity was extremely narrow or, more precisely, it was outside the limits of sensitivity of the methods employed by us.

3. STRUCTURE OF Cu_2GeS_3

We investigated in detail the structure of the compound Cu_2GeS_3. The relationship between the line intensities in the Debye powder patterns and the number of atoms per unit cell with the dimensions given in Table 2, showed clearly that the atomic distribution in unit cell of Cu_2GeS_3, Cu_2GeSe_3, or Cu_2GeTe_3 was completely analogous to that in zinc blende, with the copper and germanium atoms replaced by zinc atoms, and the sulfur atoms replaced by sulfur, selenium, or tellurium atoms.

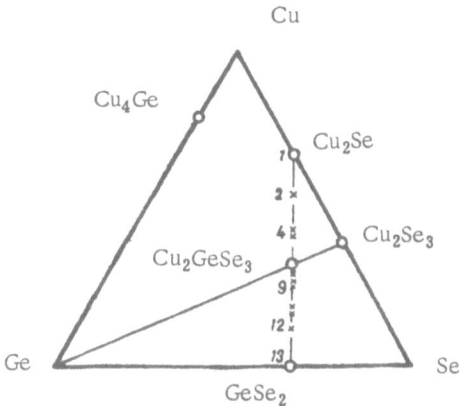

Fig. 1. Formation of the compound Cu_2GeSe_3.

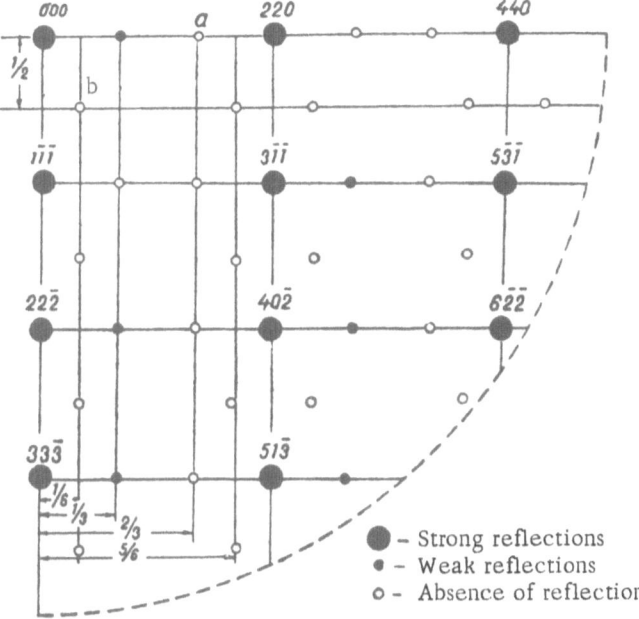

Fig. 2. Zeroth network of the reciprocal lattice of Cu_2GeS_3.

The tetragonal unit cell indicated that the distribution of the copper and germanium atoms was ordered.

Such ordering should lead to the appearance of "superstructure" lines in x-ray diffraction patterns, in which the structural amplitude should be governed by the difference of the atomic scattering factors of copper in germanium.

Since the difference between the number of electrons in copper and germanium atoms is small, the difference between the atomic factors is also small and the "superstructure" lines are not found in x-ray diffraction patterns.

Powder patterns of Cu_2GeS_3 and Cu_2GeSe_3 obtained with monochromatic radiation using prolonged exposure showed very weak lines. However, since these lines may have been due to very small amounts of impurity phases and interpretation of the powder patterns is not unambigous, they were not used to determine the "superstructure."

The "superstructure" was found by an x-ray goniometric study of a Cu_2GeS_3 single crystal. In presenting the results of this investigation, we need to use a structure, which we shall call the "pseudolattice," in which the copper and germanium atoms are not distinguished; the pseudolattice parameters listed in Table 2 will be denoted by a' and c'/2.

Our Cu_2GeS_3 crystal was a fragment of a prism with its long dimension along the [112] direction of the pseudolattice; its length was about 0.5 mm and the base measured about 0.1 mm. The x-ray diffraction patterns obtained on rotating the crystal about the [112] axis showed the absence of the "superstructure" layer lines. The development of the zeroth layer line made it possible to plot the plane network of the reciprocal lattice shown in Fig. 2.

Here, apart from "strong" reflections, which are denoted in Fig. 2 by the pseudolattice indices, there were also very weak "superstructure" reflections of two types: those located on lines joining reciprocal lattice sites in the [110] direction (one of them is denoted by "a" in Fig. 2) and those not located on such lines ("b" in Fig. 2). The developments of other layer lines showed the same distribution of reciprocal lattice sites in parallel plane networks.

The "superstructure" reflections of type "a" were less intense than those of type "b" and only some of them appeared in the developments of layer lines.

Having plotted the whole reciprocal lattice, we could easily find from the distances between the reciprocal lattice planes that the pseudolattice parameter in the [110] direction should be doubled and in the [112] direction trebled. Hence, we could find unambigously the distribution of the copper and germanium atoms in the planes parallel to (001) of the pseudolattice (Fig. 3).

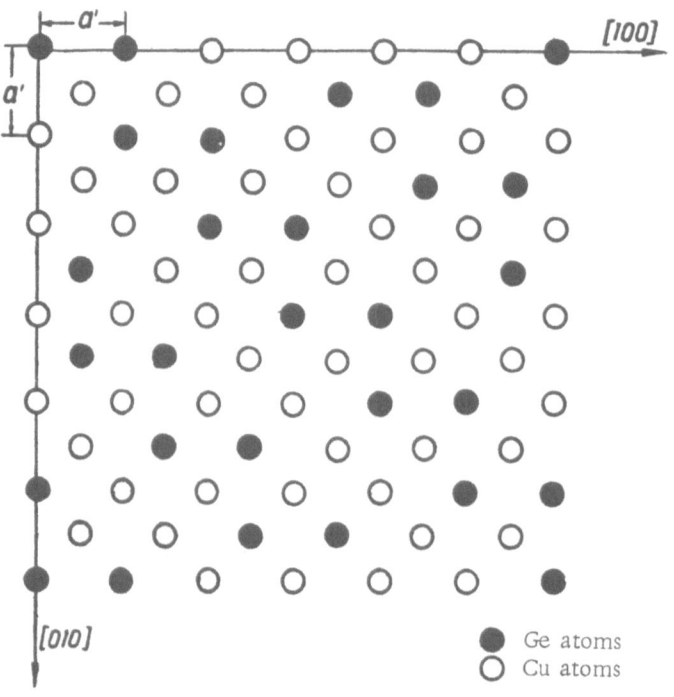

Fig. 3. Distribution of Cu and Ge atoms in the (001) plane
of the pseudolattice.

● Ge atoms
○ Cu atoms

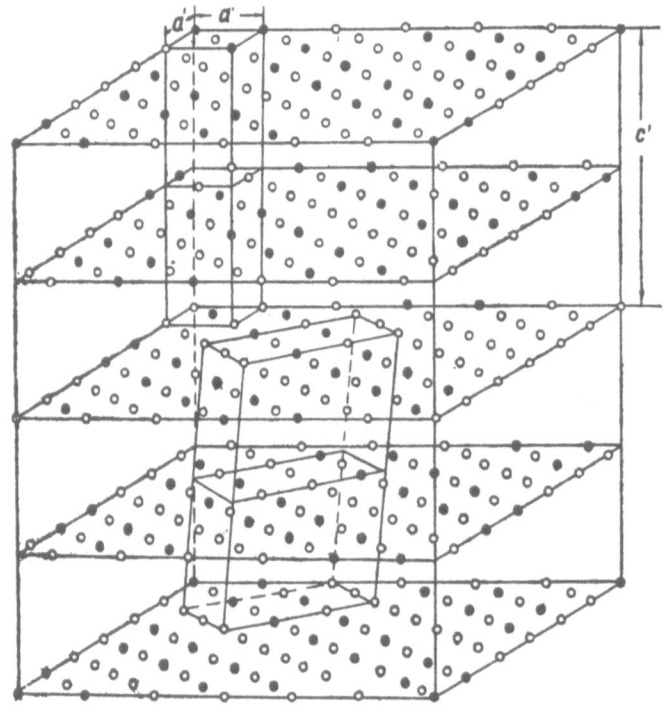

Fig. 4. Distribution of Cu and Ge atoms in Cu₂GeS₃.

The diffraction pattern described above can be obtained only for one variant of the layer distribution (Fig. 4), which can be checked by calculating the structural amplitudes.

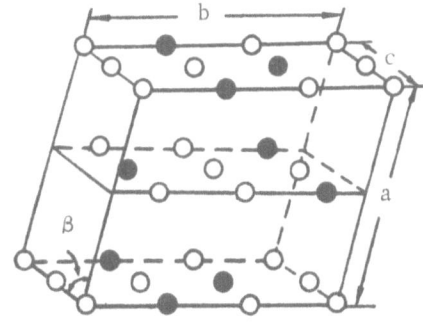

Fig. 5. Unit cell of Cu_2GeS_3 showing the distribution of Ge and Cu atoms.

On comparing structure factors, we see that for "strong" reflections the values of the structure factors will be $(4f_{Cu}+2f_{Ge}+6f_S)^2$, $(4f_{Cu}+2f_{Ge}-6f_S)^2$, or $(4f_{Cu}+2f_{Ge})^2+(6f_S)^2$, while for "weak" type "a" reflections they will be $(f_{Ge}-f_{Cu})^2$, and for "weak" type "b" reflections $3(f_{Ge}-f_{Cu})^2$.

Turning now to Fig. 4, which shows the distribution of copper and germanium atoms in Cu_2GeS_3, and taking into account the distribution of the sulfur atoms, we find that the structure has only one symmetry element: a glide reflection plane, parallel to the 110 plane of the pseudolattice. This corresponds to the space group Pc, and the unit cell dimensions are: $a = \sqrt{a'^2/2+c'^2} = 6.420$ kXU; $b = (3\sqrt{2}/2)a' = 11.277$ kXU, $c = \sqrt{2}a' = 7.518$ kXU, $\tan(180-\beta) = \sqrt{2}c'/a'$, $\beta = 125°50'$ (Figs. 4 and 5).

If we compare the ratios of the measured and calculated intensities of "strong" and "weak" reflections, we find that to calculate the intensities of "weak" and "strong" reflections we need to know very exactly the atomic factors of copper and germanium, which would have to be calculated allowing for the valence electron distribution in Cu_2GeS_3.

The absence of "very weak" reflections in Cu_2GeTe_3 may be explained by two factors. First, the relative intensity of "very weak" reflections in the case of Cu_2GeTe_3 will be less than in the case of Cu_2GeS_3 and Cu_2GeSe_3, since the composition of the first compound includes heavy atoms of tellurium, which increase the intensity of "strong" reflections. Secondly, it is possible that, in addition to the change in the relationship between the various components of chemical binding due to the replacement of lighter with heavy atoms, the distribution of the valence electrons changes, which tends to "equalize" the atomic scattering factors of copper and germanium. It is our intention to publish in the near future the results of structural investigations of the remaining compounds.

4. FORMATION OF SOLID SOLUTIONS BASED ON COMPOUNDS OF THE $A_2^I B^{IV} C_3^{VI}$ TYPE

Isovalent and heterovalent substitution in solid solutions of compounds of this type has not yet been investigated.

a) $Cu_2GeSe_3 - Cu_2SnSe_3$ System

The purpose of this work was to find whether it is possible to replace germanium with tin [6]. The results of the investigation are given in Table 3. They show that substitutional solid solutions are formed at all compositions. In addition to the data in Table 3, we shall mention some further results.

A study of the thermal conductivity showed that the alloys having compositions 2:1, 1:1, and 1:2 had conductivities of the same order as that of PbS, and that their conductivities increased with increase of the Cu_2SnSe_3 content. Measurements of the thermal conductivity were carried out immediately after synthesis. The results were:

$2Cu_2GeSe_3 - Cu_2SnSe_3 \quad \varkappa = 7.1 \times 10^{-3}$ cal·cm^{-1}·sec^{-1}·(deg C)$^{-1}$,

$Cu_2GeSe_3 - Cu_2SnSe_3 \quad \varkappa = 8.5 \times 10^{-3}$ cal·cm^{-1}·sec^{-1}·(deg C)$^{-1}$,

$Cu_2GeSe_3 - 2Cu_2SnSe_3 \quad \varkappa = 9.1 \times 10^{-3}$ cal·cm^{-1}·sec^{-1}·(deg C)$^{-1}$.

Measurements of the thermal conductivity were kindly carried out by our Czech colleague, Dr. Štourač.

b) $Cu_2GeSe_3 - Cu_2GeTe_3$ System

Three alloys, having compositions 1:3, 1:1, and 3:1, were prepared. Immediately after synthesis, the three alloys were nonequilibrium solid solutions. Annealing, carried out at 500°C for 900 hours, tendend to homogenize the alloys but its duration was insufficient. The alloys retained their tetrahedral structure.

31

TABLE 3

Compound	Structure	Lattice parameter	Melting point — crystallization temperature, °C
Cu_2GeSe_3	tetragonal	$a = 5.578 \pm 0.001$ $c = 5.474 \pm 0.001$	788
$3\ Cu_2GeSe_3 \cdot Cu_2SnSe_3$	”	$a = 5.60 \pm 0.01$ $c = 5.54 \pm 0.01$	704—765
$Cu_2GeSe_3 \cdot Cu_2SnSe_3$	ZnS	$a = 5.60 \pm 0.01$	701—769
$Cu_2GeSe_3 \cdot Cu_2SnSe_3$	”	$a = 5.61 \pm 0.01$	687—741
$Cu_2GeSe_3 \cdot 2Cu_2SnSe_3$	”	$a = 5.63 \pm 0.01$	695—744
$Cu_2GeSe_3 \cdot 3Cu_2SnSe_3$	”	$a = 5.65 \pm 0.01$	704—745
Cu_2SnSe_3	”	$a = 5.676 \pm 0.01$	709

c) $Cu_2GeSe_3 - CuGe_2Sb_3$ System

The heterovalent substitution in this system was investigated along the tie-line $Cu_2GeSe_3 - CuGe_2Sb_3$, where selenium could be substituted by antimony. The tie-line includes a hypothetical compound $CuGe_2Sb_3$, which we were unable to prepare [6].

Later studies showed that when the $CuGe_2Sb_3$ content was increased in the alloys there was a change of structure: the principal phase acquired the ZnS structure and the parameter a = 5.555 kXU.

At all points of the tie-line, the Debye powder patterns had the lines of antimony.

Beginning with the composition 1:1, germanium lines appeared on increase of the $CuGe_2Sb_3$ content. The appearance of the cubic structure with the parameter a = 5.555 kXU was due to the solution of germanium in Cu_2GeSe_3, which transformed the tetragonal into the cubic lattice [10].

Annealing for 900 hours produced no changes in the x-ray diffraction patterns and did not result in the heterovalent substitution, as erroneously suggested in [6].

d) $Cu_2GeSe_3 - GaAs$ System

The investigation of the interaction of one of the well-known and extensively investigated semiconducting compounds of the $A^{III}B^{V}$ type with Cu_2GeSe_3 offered an example of heterovalent substitution in a five-component system. As is known, GaAs crystallizes out in the ZnS lattice.

A large number of alloys was prepared. The analysis of the results obtained shows the presence of solubility in all alloys. Annealing at 500°C for 910 hours clearly increased the alloy homogeneity. The appearance of ZnS lines, beginning with 8 mol. % (GaAs) and extending up to 25 mol. % (3GaAs) with the parameter a = 5.555 kXU, indicates that GaAs which enters the Cu_2GeSe_3 lattice disturbs the ordering of the Cu and Ge atomic distribution. A more detailed presentation of these results will be published later.

5. DISCUSSION OF RESULTS

Comparing the published data with our results, we note the following points.

Table 2 lists the values of the parameters "a" and "c" for compounds containing copper and germanium; these parameters were obtained from x-ray diffraction patterns and the ration c/a was close to unity.

The numerical values of the parameters obtained by us (Table 2) agree with those reported by Hahn [2] but we cannot concur with his structure determinations because the compositions of our compounds were not analogous to the composition of chalcopyrite $CuFeS_2$.

Some workers [8, 9] are of the opinion that compounds containing copper and germanium have the ZnS-type structure, and that the compound Cu_2GeSe_3 has the parameter a = 5.55 A.

This disagreement between our data and 'theirs may be due to the excess of Ge in the samples used by those workers [8, 9] since the parameter reported by them corresponds to the parameter of the phase obtained on disolving germanium in Cu_2GeSe_3 [10].

The authors of [8] assume that one can use Magnus's criterion (the ratio of ionic radii of the cation and anion lying in the range from 0.225 to 0.415) to predict the existence of ternary compounds. In the opinion of these workers [8] the compound Cu_2SnTe_3 exists because $r_{Sn4+}/r_{Te2-} = 0.36$ and the compound Cu_2PbTe_3 does not exist because $r_{Pb4+}/r_{Te2-} = 0.41$.

However, the following observations can be made about this conclusion. If we agree with the use of ionic radii, and this is controversial [11], it is not clear why they have not used ionic radii of group I elements, for example, copper, which has a considerably larger ionic radius than group IV elements. Moreover, the ratios of the ionic radii for the $A_2^I B^{IV} C_3^{VI}$ compounds which are known to exist do not all lie withing the limits 0.415-0.225. Finally, if we bear in mind that in $A_2^I B^{IV} C^{VI}$ compounds there are twice as many A^I atoms as B^{IV} atoms, the conclusions of the authors of [8] become completely incomprehensible.

The problem of the real existence of ternary compounds with a tetrahedral distribution of atoms cannot be solved using ionic radii alone.

If we compare the chalcopyrite-type structure with the structure of Cu_2GeS_3, we readily find that in the former the ordering leads to the formation of atomic sequences in which copper and iron atoms alternate (for example, along [001]),or there are alternate series of like atoms (for example, along [100]). In Cu_2GeS_3, there are series in which two atoms of copper alternate with one atom of germanium (for example, along [010]),or sequences with two series of alternating copper and germanium atoms and one series of copper atoms (for example, along [001]).

Isovalent and heterovalent substitution of group IV or group VI elements is possible in $A_2^I B^{IV} C_3^{VI}$ compounds, as in $A^{III} B^V$ compounds. Disordering occurs when solid solutions are formed as a result of which the lattice becomes cubic.

The majority of substances of the $A_2^I B^{IV} C_3^{VI}$ group investigated by us were photosensitive in thin-film form, which may be of some practical interest.

Further work is needed on the main physico-chemical and electrical properties of substances of this type.

LITERATURE CITED

1. C. H. L. Goodman, Semiconducting Substances (Collection of Papers) [Russian translation], IL (1960).
2. H. Hahn, Proceedings of the Seventeenth Congress on Applied Chemistry, Munich (1959).
3. O. G. Folberth and H. Pfister, Halbleiter und Phosphore (1958).
4. N. A. Goryunova and V. I. Sokolova, Izv. Moldavsk. Filiala Akad. Nauk SSSR, No. 3(69) (1960).
5. N. A. Goryunova, Izv. Moldavsk. Filiala Akad. Nauk.SSSR, No. 3(69) (1960).
6. N. A. Goryunova, G. K. Averkieva, and Yu. V. Alekseev, Izv. Moldavsk. Filiala Adad. Nauk SSSR, No. 3(69) (1960).
7. N. A. Goryunova, Vestn. Leningr. Gos. Univ. im. A. A. Zhdanova, No. 10 (1961).
8. L. S. Palatnik, Yu. F. Komnik, V. M. Koshkin, and E. K. Belova, Dokl. Akad. Nauk SSSR, 137(1) (1961).
9. L. S. Palatnik, Yu. F. Komnik, E. K. Belova, and L. V. Atroshchenko, Kristallografiya 6(6) : 960 (1961).
10. N. A. Goryunova and Chiang Ping-hsi, Abstracts of Papers Presented at the All-Union Conference on Semiconducting Compounds, Leningrad (1961).
11. N. A. Goryunova, Doctoral dissertation, Moscow (1958).

THE CARRIER MOBILITY AND EFFECTIVE MASS IN GLASSY

$Tl_2SeAs_2Te_3$

A. M. Andriesh and B. T. Kolomiets

Quite a few papers have been published on glassy semiconductors. However, there has been practically no investigation of the behavior of charge carriers in these materials. This is because the low values of the carrier moblity in chalcogenide glasses [1] make it difficult to apply galvanomagnetic and thermomagnetic methods, which are so successful in the study of crystalline semiconductors.

To find information on the behavior of carriers in glassy semiconductors, it is necessary to know the electrical conductivity, the thermoelectric power, the ratio of the mobilities, and the effective carrier mass. The present work reports the results of measurements of the temperature dependence of the electrical conductivity and the thermoelectric power of glassy $Tl_2SeAs_2Te_3$. The ratio of the mobilities and the effective carrier mass were estimated from these measurements.

The present test material was selected because it has a higher conductivity ($\approx 10^{-3}$ ohm$^{-1} \cdot$ cm^{-1}) than less complex chalcogenide glasses and, therefore, is a very convenient material for the study of the electrical conductivity and the thermoelectric power over a relatively wide range of temperatures.

INVESTIGATION METHOD

The alloy $Tl_2SeAs_2Te_3$ was prepared from its separate components in evacuated quartz ampoules [2].

The absence of a crystalline phase in the test material was established by X-ray diffraction analysis. Investigations of the microstructure and microhardness showed that the alloy $Tl_2SeAs_2Te_3$ was homogeneous. The homogeneity was checked also by measurements of the conductivity and of the sign of the thermoelectric power along the sample length.

The test samples were cut from an ingot in the form of parallelepipeds of the following approximate dimensions: $3 \times 3 \times 10$ mm. Colloidal graphite electrodes were used in the measurements of the electrical conductivity. The ohmic nature of the contacts was checked by measuring the current—voltage characteristics, which were found to be linear up to fields of ≈ 5 V/cm. In stronger fields, the current—voltage characteristics became nonlinear due to the Joule heating of the sample.

The electrical conductivity and thermoelectric power were measured with a calibrated electrometer amplifier having an input impedance of 10^{11} ohm. To measure the thermoelectric power, an instrument of the Emel'yanenko and Trishin type was constructed, which also allowed several electrical properties to be measured simultaneously [3].

The thermoelectric power was measured using temperature differences of 10-20°C between opposite ends of the sample, but it was found to be independent of these temperature differences.

The range of temperatures employed in the study of the electrical conductivity and thermoelectric power of $Tl_2SeAs_2Te_3$ was governed, on the one hand, by the temperature at which this substance softened ($\sim +70$°C) and, on the other, by the limitations of the measuring circuit. The maximum resistance which we

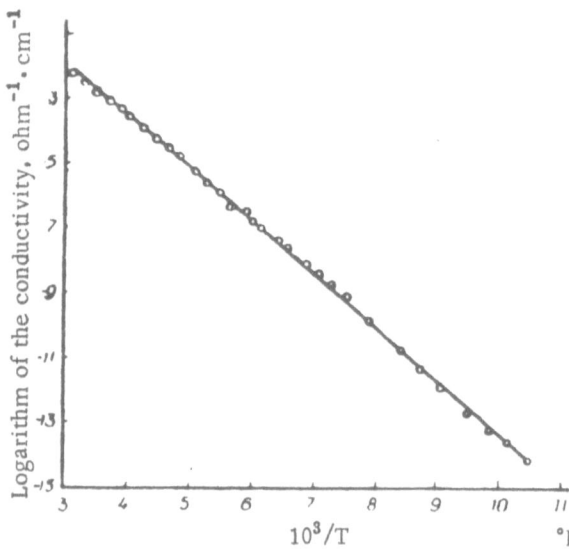

Fig. 1. Temperature dependence of the electrical conductivity of $Tl_2SeAs_2Te_3$.

could measure with the electrometer amplifier was of the order of 10^{13}-10^{14} ohm. Since the thermoelectric power was measured under open-circuit conditions, we could only measure the thermoelectric power of samples whose resistance was not higher than 10^8-10^9 ohm. The samples had this resistance on cooling to $\sim -110°C$.

RESULTS OF THE MEASUREMENTS AND DISCUSSION

Our measurements showed that the electrical conductivity of $Tl_2SeAs_2Te_3$ at room temperature was 2.5×10^{-3} $ohm^{-1} \cdot cm^{-1}$. The dark conductivity decreased exponentially on cooling to liquid-nitrogen temperature. A plot of the dependence log $\sigma = f(10^3/T)$ for one of the $Tl_2SeAs_2Te_3$ samples is shown in Fig. 1. This indicates that there is only one slope, corresponding to 0.67 eV, throughout the whole range of tests temperatures.

At room temperature, the alloy had a considerable thermoelectric power: 930 μV/deg. The sign of the thermoelectric power represented p-type conduction at all temperatures. On cooling the samples, their thermoelectric power rose, reaching a value of 1550 μV/ deg at $-100°C$. The temperature dependence of the thermoelectric power is shown in Fig. 2a. It was also interesting to plot the dependence of the thermoelectric power on the reciprocal of temperature: this is shown in Fig. 2b, where this dependence is a straight line with a slope corresponding to 0.26 eV.

Kolomiets and Nazarova [4, 5] reported that the impurity conduction did not appear in chalcogenide glasses.

The fact that the values of the activation energy of chalcogenide glasses, determined from the temperature dependence of the dark conductivity, from the optical absorption edge, and from the photoconductivity, are practically the same [4, 6] indicates that chalcogenide glasses are intrinsic semiconductors. The alloy $Tl_2SeAs_2Te_3$ investigated by us is no exception. Its intrinsic conductivity is confirmed by the single slope of the dependence of the logarithm of the dark conductivity on temperature.

Since the thermoelectric power was measured in the intrinsic conduction region, we could, as for crystal semiconductors, determine the ratio of the carrier mobilities from the slope of the dependence $\alpha = f(1/T)$ and from the known value of the forbidden band width.

The possibility of using this method of determining the mobilities in glassy semiconductors follows from Gubanov's theory of amorphous semiconductors [7].

Using one-dimensional and three-dimensional models of a liquid, Gubanov showed that the energy band structure is retained when the long-range order disappears [8, 9]. Allowing for liquid and phonon scattering, Gubanov calculated the temperature dependence of the transport coefficients for glassy semiconductors [7]. He obtained the following expression for the thermoelectric power when only one type of carrier is present:

$$\alpha = -\frac{k}{e}\left[\frac{\mu}{kT} - 2\frac{A' + (A'' - B')\frac{T}{\theta}}{A' + A''\frac{T}{\theta}}\right],\tag{1}$$

where μ is the Fermi level, T the temperature, k Boltzmann's constant, A' and B' are temperature-independent coefficients representing the "liquid" scattering of carriers, A'' is a coefficient representing the scattering of electrons on thermal vibrations, and Θ the Debye temperature.

36

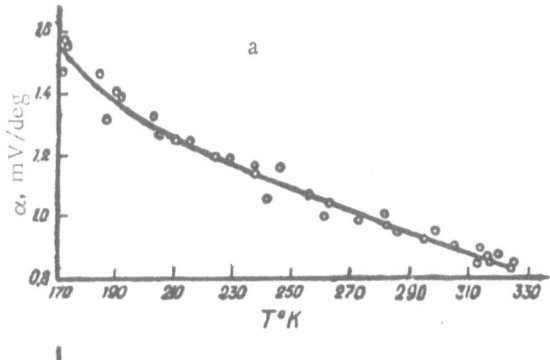

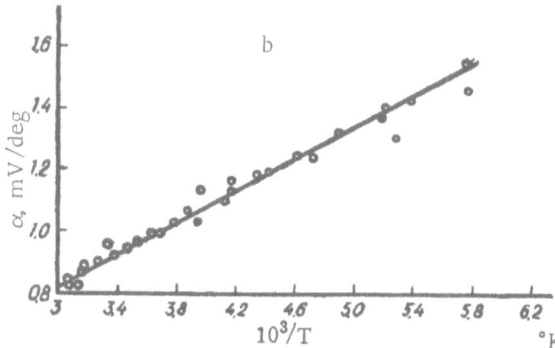

Fig. 2. Temperature dependence of the thermoelectric power of $Tl_2SeAs_2Te_3$.

Formula (1) has the same form as the expression for the thermoelectric power of crystalline semiconduunctors. However, in contrast to crystals, the term depending on the nature of scattering has coefficients representing phonon and liquid scattering of carriers. If we assume that the nature of hole and electron scattering in glassy semiconductors is the same, it follows that the expression for the thermoelectric power in the intrinsic conduction region obtained for crystals applies also to glassy semiconductors:

$$\alpha = -\frac{k}{e}\left[\frac{u_n - u_p}{u_n + u_p}\left(\frac{E_0}{2kT} + A - \frac{a}{2k}\right) - \frac{3}{4}\ln\frac{m_p^*}{m_n^*}\right],$$ (2)

where u_n, u_p are, respectively, the electron and hole mobilities; E_0 is the forbidden band width at absolute zero; a is a coefficient which occurs in the exponent of the temperature dependence of E_0; m_n^*, m_p^* are, respectively, the electron and hole masses. The symbol A in Eq. (2) represents:

$$A = 2\frac{A' + (A'' - B')\frac{T}{\theta}}{A' + A''\frac{T}{\theta}}.$$ (3)

To determine the ratio of the carrier mobilities, we shall assume that it is independent of temperature in the region investigated by us. Moreover, we must check that the temperature dependence of the thermoelectric power is not affected by the temperature dependence of A. If the phonon scattering predominates over the liquid scattering $[A'' (T/\theta) \gg A', B' (T/\theta)]$, then A = 2. If the liquid scattering predominates over the phonon scattering $[A' (T/\theta) \ll A', B' (T/\theta)]$, A assumes the following value:

$$A = 2 - \frac{B'}{\theta A'} T.$$ (4)

Thus, in the phonon-scattering case, the temperature dependence of the thermoelectric power is determined completely by the first term of equation (2), i. e., by $\frac{k}{e}\frac{u_n - u_p}{u_n + u_p}\frac{E_0}{2kT}$. In the liquid-scattering case, the temperature dependence of the thermoelectric power may be given by:

$$\alpha = -\frac{k}{e}\left[\frac{u_n - u_p}{u_n + u_p}\left(\frac{E_0}{2kT} + 2 - \frac{B'}{\theta A'} T - \frac{a}{2k}\right) - \frac{3}{4}\ln\frac{m_p^*}{m_n^*}\right].$$ (5)

It follows from Fig. 2b that the temperature dependence of the thermoelectric power of $Tl_2SeAs_2Te_3$ is of the $\alpha = (C_1/T) + C_2$ type, where C_1 and C_2 are coefficients which are independent of temperature.

Comparison of Fig. 2b with Eq. (5) leads us to the conclusion that, in the investigated range of temperatures, either the phonon scattering dominates over the liquid scattering in $Tl_2SeAs_2Te_3$ or $(B'/\theta A')T \ll E_0/2kT$, so that the temperature dependence of the thermoelectric power is determined only by the term $\frac{k}{e}\frac{u_n - u_p}{u_n + u_p}\frac{E_0}{2kT}$.

These considerations allow us to ignore the temperature dependence of A and to determine from equation (2) the carrier mobility ratio in $Tl_2SeAs_2Te_3$.

The ratio determined in this way was found to be 0.13. This means that the hole mobility is one order of magnitude greater than the electron mobility.

It is interesting to estimate the value of the effective hole mass in $Tl_2SeAs_2Te_3$. Bearing in mind that the mobility ratio is 0.13, we can then use approximately the expression for the thermoelectric power in the case of one type of carrier [Eq. (1)]. The value of the carrier density in $Tl_2SeAs_2Te_3$ at room temperature, equal to 6×10^{17} cm^{-3}, was taken from the work of Kolomiets and Nazarova [1]. Since we had no data for the determination of the coefficients representing the liquid scattering, we estimated the effective hole mass in $Tl_2SeAs_2Te_3$ on the assumption that the phonon scattering was the dominant process. The value of the effective hole mass found with these assumptions was of the order of $30m_0$, where m_0 is the free electron mass.

Such a high value of the effective mass seems, in our opinion, to be in agreement with the low mobility of carriers in $Tl_2SeAs_2Te_3$ which, according to Kolomiets and Nazarova [1], amounts to 10^{-2} cm$^2 \cdot$V$^{-1} \cdot$sec^{-1}.

CONCLUSIONS

In applying to glassy substances the method of determining the carrier mobility ratio in crystalline semiconductors, we have come to the following conclusions:

1) the hole mobility in glassy $Tl_2SeAs_2Te_3$ is approximately one order of magnitude higher than the electron mobility;

2) the effective hole mass is of the order of $30m_0$.

The x-ray structural analysis of $Tl_2SeAs_2Te_3$ was carried out by A. A. Vaipolin, and the microstructure analysis by V. P. Shilo, and the authors are grateful to them for this help.

The authors are also indebted to O. V. Emel'yanenko for his valuable advice.

LITERATURE CITED

1. B. T. Kolomiets and T. F. Nazarova, Fiz. Tverd. Tela 2:395 (1960).
2. N. A. Goryunova and B. T. Kolomiets, Zh. Tekhn. Fiz. 21:984 (1955).
3. O. V. Emel'yanenko and N. V. Trishin, PTÉ No. 1, 98 (1960).
4. B. T. Kolomiets and T. F. Nazarova, Fiz. Tverd. Tela, Sbornik 2:22 (1955).
5. B. T. Kolomiets and T. F. Nazarova, Fiz. Tverd. Tela 2:174 (1960).
6. T. N. Vengel' and B. T. Kolomiets, Zh. Tekhn. Fiz. 27:2484 (1957).
7. A. I. Gubanov, Zh. Tekhn. Fiz. 27:3 (1957).
8. A. I. Gubanov, Zh. Eksperim. i Teor. Fiz. 26(2) (1954).
9. A. I. Gubanov, Zh. Eksperim. i Teor. Fiz. 28:401 (1955).

INVESTIGATION OF THE MICROHARDNESS ANISOTROPY
OF TELLURIUM

Yu. S. Boyarskaya, V. N. Lange, and T. I. Lange

The elemental crystalline semiconductor, tellurium, is of special interest because of its charateristic structure. Its lattice, which is of the hexagonal type (D_3 class), consists of parallel helical "infinite" chains, directed along the C axis [1]. The projections of the chains on the basal plane are equilateral triangles located at the corners and in the center of a regular hexagon, as shown in Fig. 1.

Neighboring atoms in each chain are bound by strong covalent bonds, while the forces between the chains are weak, being partly of the van der Waals type and partly of a metallic nature [2]. In accordance with this nature of the forces and the structure, the cleavage planes ($10\bar{1}0$) are also glide planes [3].

The strong bond anisotropy in the lattice makes many of the physical properties strongly anisotropic [4]. However, until recently, there has been little research on this aspect. The anisotropy of the mechanical properties has been practically neglected [5, 6].

In view of this, we decided that a study of the microhardness anisotropy of tellurium would be of interest. The results obtained in this study are reported below.

EXPERIMENTAL METHOD

All the tests were made on single-crystal samples prepared from triply vacuum-distilled tellurium containing about 10^{15} cm^{-3} active uncompensated impurities, as found from the value of the Hall coefficient at -183°C.

Single crystals were grown by the Bridgman method [7], at about 10 cm/hour, in glass ampoules first evacuated and then filled with spectroscopically-pure argon. The final single crystals were in the form of cylinders of 3-4 mm diameter and 100 mm long. The C axis was usually along the cylinder axis or made a small angle with the latter.

To study the microhardness along the ($10\bar{1}0$) face, the crystal was cleaned by a light application of a safety-razor blade edge. In this case, there was no need for additional treatment of the surface.

The direction of the C axis was determined first as a line of intersection of two cleavage planes in order to obtain the (0001) face. The crystal was then cut with a diamond saw at right-angles to the C axis and the resultant plane was polished mechanically until a mirror-like surface was achieved.

The microhardness was measured with a type PMT-3 instrument. A simple device was attached to the object stage of PMT-3 for rotating the sample about the optical axis of the instrument. The microhardness anisotropy was investigated by the scratch method [8], the applicability of which to our purpose was proved in [9].

The surface relief of a sample near indentations and scratches was examined with an interference microscope of the MII-4 type.

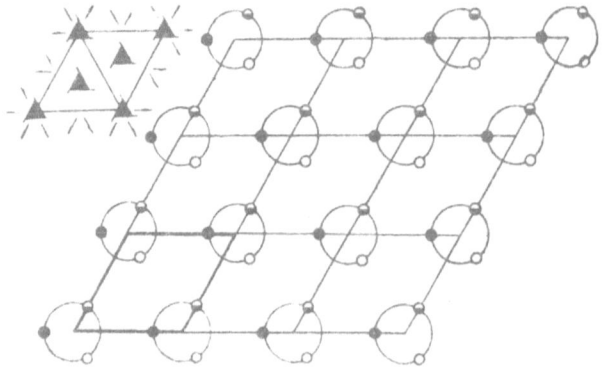

Fig. 1. Projection of the crystal lattice of tellurium on the basal plane. Atoms at various heights are denoted, respectively, by open, black, and half-black circles. The thick lines define a unit cell. The crystal-symmetry elements are shown in the top left-hand corner of the figure.

a

b

Fig. 2. Indentation shape of the (10$\bar{1}$0) face of tellurium: a) rhomboidal indentations, and b) butterfly shaped. The arrow shows the C-axis direction (this applies also to subsequent figures). Slip bands are clearly visible.

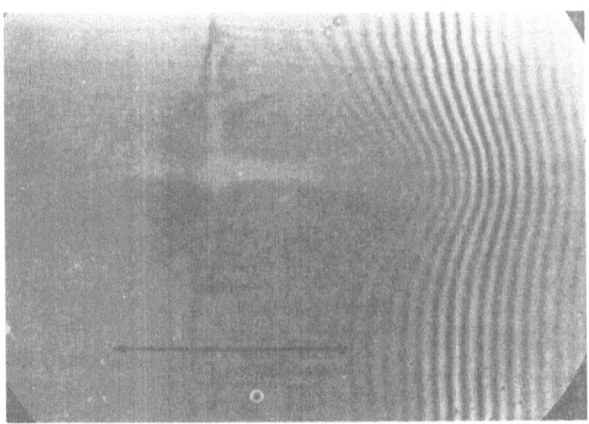

Fig. 3. Bending of interference bands near
a rhomboidal indentation.

EXPERIMENTAL RESULTS AND DISCUSSION

A. (10$\bar{1}$0) Face

As mentioned in earlier work [6], indentations on this face have an unusual shape, very different from a square. In the present work, we carried out a detailed study of the dependence of the indentation shape on the orientation. The following results were obtained.

If a sample was oriented so that one of the diagonals of the indentation made by a pyramid coincided with the C axis, the indentation was in the form of a rhombus (Fig. 2a). The long diagonal of this rhombus was perpendicular to the C axis and had sharp boundaries, while the ends of the short diagonal, along the C axis, were not sharp.

If the indentation diagonal made an angle of 45° with the C axis, the indentations assumed a butterfly shape (Fig. 2b). At angles between 0 and 45° with the C axis, the indentation shape was intermediate between the rhomboidal and butterfly shape.

A study of the surface relief near indentations, using a microinterferometer, showed a bending of the interference bands along the C axis direction at quite a large distance from the center of the indentation; the bending increased on approach to the indentation. This was observed both for rhomboidal and butterfly-shaped indentations (Fig. 3). The direction of bending of the interference bands indicated the presence of a valley near the indentation, running from the middle of the indentation along the C axis. We concluded from this that an indentation made with a pyramid depressed the surface along the C axis, perhaps breaking the chains.

If this is correct, the angle of the depression should be greater than the angle of the point of the diamond pyramid. From the bending of the interference bands, we deduced the depression of the surface for indentations of two orientations (C axis making 0° and 45° with the indentation diagonal).

The results are listed in Table 1. In this table, d represents the indentation diagonal (the long diagonal for rhomboidal indentations but any diagonal for butterfly-shaped ones), l_0 is the distance from the center of the indentation to a point where the considerable bending of interference bands becomes noticeable, l is the distance from the center of the indentation of a point where the depth of the surface depression is h, and α is the angle of this surface depression (Fig. 4), found from the formula

$$\tan \alpha = \frac{l_0 - l}{h}.$$

TABLE 1

Indentation shape	Load, g	d, μ	l_0, μ	l, μ	h, μ	$\alpha°$
Rhomboidal	2	11	18	13	0.18	88
	5	15	24	18	0.21	88
	10	15	28	18	0.20	89
	20	22	30	27	0.27	85
	100	49	78	52	0.76	88
Butterfly shaped	2	14	26	20	0.21	88
	5	21	36	30	0.24	88
	10	28	57	41	0.35	88
	20	41	80	56	0.49	89
	100	92	152	87	0.10	89

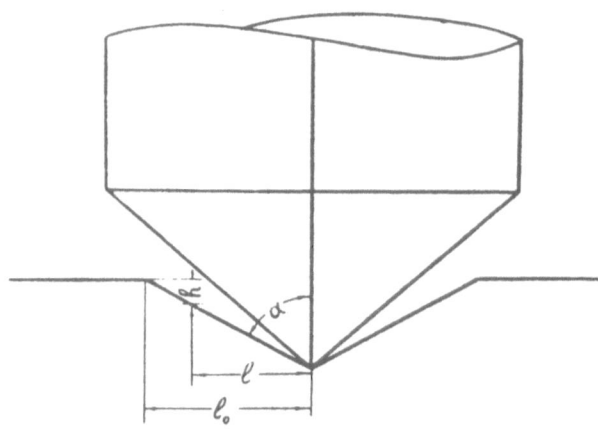

Fig. 4. Determination of the angle of depression.

It follows from Table 1 that, in all cases, considerable depression of the surface began at distances about 1.5-2 times greater than the indentation diagonal. The angle of depression was approximately the same for indentations of all dimensions and, in accordance with the hypothesis put forward above, it was larger than the half-angle (74°) between the edges of a standard diamond pyramid of the indenter.

Thus, we may conclude that the shape of the indentations on the (10$\bar{1}$0) face is affected in a decisive manner by the depression of the surface along the C axis. This is of great interest because until now it has been assumed that the indentation shape is affected mainly by two factors: elastic recovery and a pile-up of the material at the indentation edges [10-12].

We must note that, because of this characteristic shape of the indentations on the (10$\bar{1}$0) face of tellurium, we could not decuce the microhardness from the indentation by means of the formula

$$H = 1854 \cdot \frac{P}{d^2}.$$

TABLE 2

Scratch direction	0°	15°	30°	45°	60°	75°	90°
H, kg/mm²	25,3	23.8	22.1	17.8	15.1	14.9	15.9

Scratch direction	105°	120	135°	150°	165°	180°	
H, kg/mm²	15,1	14.9	16.9	20.1	22.9	25.3	

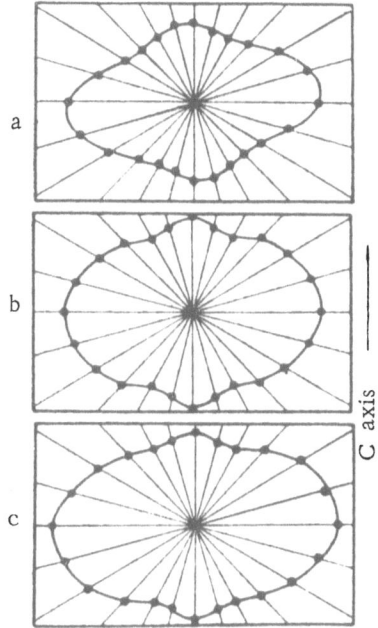

Fig. 5. Hardness rosettes for the (10$\bar{1}$0) face of tellurium obtained with a pyramid loaded with 3 g ("b" and "c") and 5 g ("a"). The "c" rosette is plotted from the data in Table 2.

Therefore, we used the scratch method to find the hardness of this face. The scratching was carried out with a diamond pyramid under loads of 3, 5, and 10 g, using the same instrument, PMT-3. The scratches were made every 15°, 5-10 scratches for each direction. Each scratch was measured in several places. The hardness was deduced from

$$H = \frac{P}{d^2},$$

where P is the load on the pyramid and d is the average scratch width.

The results of measurements for one of the samples are listed in Table 2. The measurements were carried out at P = 3 g. The error in measurement of the hardness amounted to 1-1.5 kg/mm², so that the observed hardness anisotropy could not be due to the experimental error.

Results similar to those listed in Table 2 were obtained for the other five test samples.

From these measurements, we plotted hardness rosettes, three of which are shown in Fig. 5.

An inspection of these rosettes shows that the hardness along the C axis direction was lower than at right-angles to this direction, in good agreement with the bond anisotropy in tellurium.

The anisotropy of the mechanical properties of the (10$\bar{1}$0) face also affected the shape of the scratches along different directions. The scratches along the C-axis had even edges, while those at right-angles were tree-like in shape. Scratches along intermediate directions had one side of the groove narrower than the other. All these observations are illustrated in Fig. 6.

A study of the surface relief near scratches of various orientations showed that near the scratches along the C axis the pile-up of material was either absent or slight, while in the case of scratches at right-angles to the C axis, large pile-ups were observed (Fig. 7). The scratches along the C axis had depressions at the beginning and end of each scratch; in the case of scratches at right-angles to the C axis, pile-ups were observed at the scratch ends. It is also interesting that the depth of a scratch parallel to the C axis was considerably greater than that of one at right-angles to that axis (Fig. 7). Measurements showed that the depth of the scratches at right-angles to the C axis was considerably less than that expected from the width of the scratch and the geometrical dimensions of the diamond pyramid.

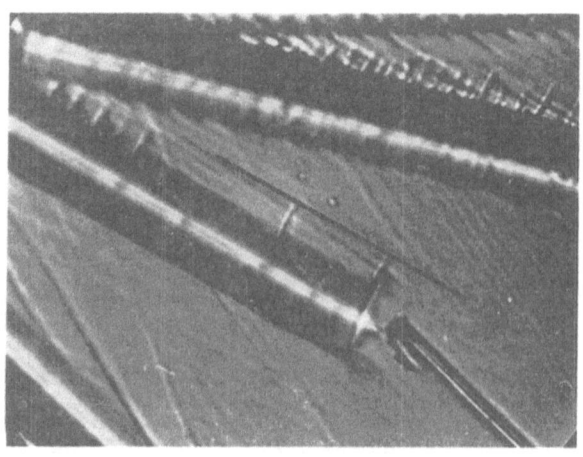

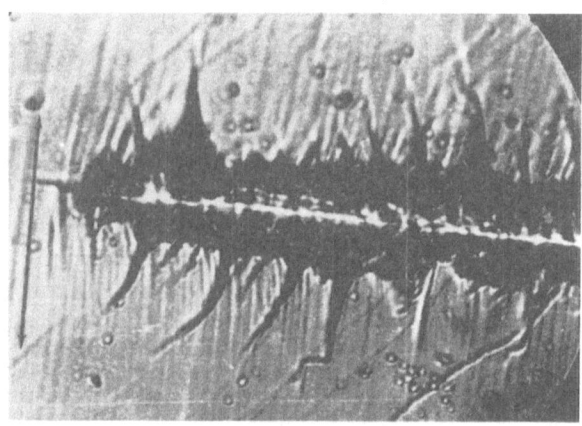

Fig. 6. Shape of scratches of the (10$\bar{1}$0) face of tellurium.

TABLE 3

Scratch direction	0°	30°	60°	90°	120°	150°	180°	210°	240°	270°	300°	330°	360°
H, kg/mm²	61.5	61.5	71.8	71.8	55.9	64.6	78,6	58.6	53.4	61.5	78.6	64.6	58.6

These observations led us to the following conclusion about the mechanism of scratching. If a scratch is made along the C axis, it is preceded by a depression, the scratching is relatively easy, and only small pile-ups form at the sides of the scratches. When scratching is carried out at right-angles to the C axis, at right-angles to the direction of strong bonds, the scratch is preceded by a pile-up which impedes the motion of the pyramid; large pile-ups appear also at the sides of the scratch. It is possible also that this accounts for the anomalous shallowness of the scratch.

a

b

Fig. 7. Bending of interference bands at scratches;
a) parallel to the C axis and b) perpendicular to the C axis.

Fig. 8. Shape of indentations on the (0001) face of tellurium.

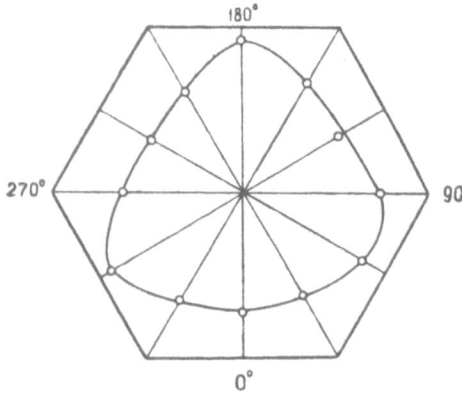

Fig. 9. Hardness rosette for the (0001) face of tellurium.

Near indentations and scratches slip bands were observed along the C-axis direction (Fig. 6a), in agreement with the published work [5, 13]. However, in addition to these bands, we sometimes observed other slip bands, making an angle of about 30° with the C axis (Fig. 6a). It is interesting to note that the minimum value of the scratch hardness lay approximately in the same direction.

B. (0001) Face

The shape of the indentations on this face was normal (Fig. 8). The microhardness calculated from the value of the diagonal was 90 kg/mm^2, with a maximum possible error of ±1.5 kg/mm^2.

The scratch hardness, as for the $(10\bar{1}0)$ face, was found to depend on the direction of scratching. The results obtained in one particular test, using a pyramid loaded with 5 g, are listed in Table 3.

In tests on the scratching of the (0001) face, the zeroth direction was assumed to be the $[10\bar{1}0]$ direction.

Results similar to those listed in Table 3 were obtained also for the other test samples (five samples were investigated). Table 3 shows that the hardness of the (0001) face was considerably higher than that of the $(10\bar{1}0)$ face (cf. Table 2).

Using the data listed in Table 3, we plotted the hardness rosette for the (0001) face, shown in Fig. 9.

Since the errors in the measurements of the hardness on the (0001) face did not exceed 2-3 kg/mm^2, we may conclude from the results in Table 3 and Fig. 9 that the hardness of this face is also anisotropic. Examination of Fig. 9 also shows that the dependence of the microhardness on the direction as a three-fold symmetry axis, like the symmetry of the (0001) face (Fig. 1), and does not contradict Neumann's principle [14]. Similar rosettes were observed earlier for the (111) faces of various cubic crystals [9].

It should be noted that the hardness depends on the direction, i.e., in the forward and reverse sense of scratching along a given line. Boyarskaya [9] called this the polar anisotropy.

In some samples, we obtained approximately the same value of hardness on scratching in two opposite directions along those angles for which the highest polar anisotropy should be observed (i.e., the angles 0-180°, 60-240°, 120-300°). In our opinion, this is not of basic importance and was probably due to departures of the face from strict perpendicularity to the C axis.

Investigation of the interference patterns near scratches made along the same line but in opposite directions showed that they were of different widths and that in front of narrow scratches there was a pile-up of the material, while in front of the broad scratches there was a depression. Thus, the polarity of the scratching appeared not only in the scratch width but also in the deformation preceding it.

It is interesting to note that this type of polarity is also observed in those cases where the scratch widths are indistinguishable. The fact that there was a pile-up of material in front of narrow scratches and depressions in front of broad ones has been noted already for the $(10\bar{1}0)$ face. However, for scratches on the $(10\bar{1}0)$ face either a pile-up or a depression was observed at both ends of the same scratch. In the case of scratches on the (0001) face, a pile-up at one end of a scratch was accompanied by a depression at the other end, and conversely.

The surface near scratches on the (0001) face was either level or had small pile-ups (Fig. 10). The shapes of the scratches along various angular directions were approximately the same. The scratch edges were usually straight but sometimes a splitting of the edge was observed (Fig. 11). Slips bands were not observed near the indentations and scratches on the (0001) face, which is in agreement with the observation that the slip in tellurium proceeds along the $(10\bar{1}0)$ planes along the [1120] directions [5].

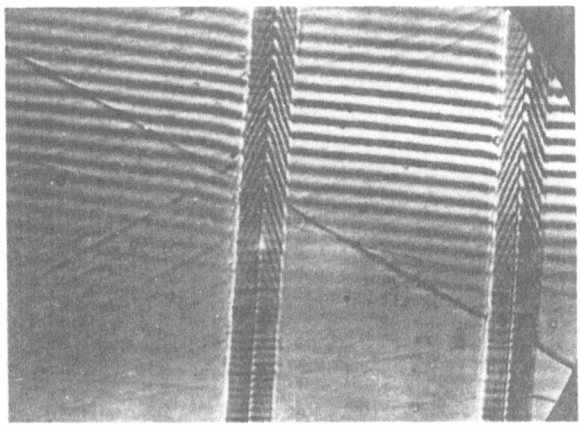

Fig. 10. Bending of interference bands at scratches on the
(0001) face of tellurium.

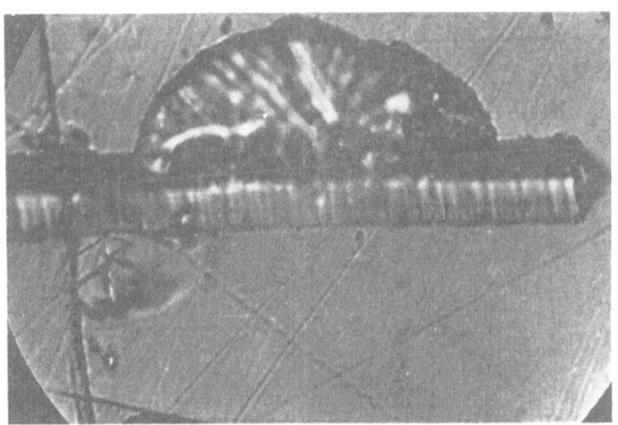

Fig. 11. Shape of a scratch on the (0001) face of
tellurium. A large chipped area is clearly visible.

CONCLUSIONS

1. Indentations on the (10$\bar{1}$0) face of tellurium are of unusual shape: rhomboidal or butterfly. This shape is due to the depression of the surface along the C axis.

2. Hardness rosettes for the (10$\bar{1}$0) face show, in agreement with the bond anisotropy in tellurium, that the hardness along the C axis is lower than that at right-angles to that axis.

3. Apart from the hardness anisotropy, the (10$\bar{1}$0) face of tellurium exhibits anisotropy of the scratch shape.

4. The (0001) face of tellurium exhibits anisotropy of the scratch hardness, the shape of the hardness rosettes being in agreement with the symmetry of this face.

5. The scratch hardness of the (0001) face of tellurium is considerably higher than that of the (10$\bar{1}$0) face.

6. Slip bands were observed on the (10$\bar{1}$0) face; they were directed at about 30° to the C axis.

LITERATURE CITED

1. B. F. Ormont, Structure of Inorganic Substances, GITTL (1950).
2. A. von Hippel, J. Chem. Phys. 16:4 (1948).
3. E. Schmid and G. Wasserman, Z. Physik 46:653 (1927).
4. V. P. Zhuze, Semiconducting Materials, Leningrad (1957).
5. R. J. Stokes, T. L. Johnston, and C. H. Li, Acta Met. 9:45 (1961).
6. V. N. Lange and A. R. Regel', Fiz. Tverd. Tela 1:559 (1959).
7. H. E. Buckley, Crystal Growth [Russian translation], IL (1954).
8. V. D. Kuznetsov, Solid-State Physics, Vol. I, Tomsk (1937).
9. Yu. S. Boyarskaya, Uch. Zap. Kishinevsk. Univ. 17:159 (1955), Yu. S. Boyarskaya and R. P. Rutskaya, Uch. Zap. Kishinevsk. Univ. 55:67 (1960), Yu. S. Boyarskaya, Kristallografiya 2(5):709, (1957); Yu. S. Boyarskaya and M. I. Val'kovskaya, Kristallografiya 7(2):261, (1962).
10. Yu. S. Boyarskaya, M. I. Val'kovskaya, and B. S. Tsukerblat, Uch. Zap. Kishinevsk. Gos. Univ. 49:32 (1961).
11. V. M. Glazov and V. N. Vigdorovich, Microhardness of Metals, Metallurgizdat (1962).
12. B. W. Mott, Micro-Indentation Hardness Testing [Russian translation], Moscow (1960).
13. J. S. Blakemore, J. W. Schulz, and K. C. Nomura, J. Appl. Phys. 31:2226 (1960).
14. J. F. Nye, Physical Properties of Crystals [Russian translation], IL (1960).

SOLID SOLUTIONS IN THE $Ga_2Se_3 - In_2Se_3$ SYSTEM

A. A. Vaipolin and V. S. Grigor'eva

In recent years, defect semiconducting compounds represented by the general formula $A_2^{III}B_3^{VI}$ have attracted more and more attention. These compounds have certain characteristic features, interesting not only from the point of view of their electrical properties but also from the point of view of the interactions between the various compounds involved.

It has been found that ternary solid solutions with wide ranges of homogeneity can be formed from such compounds [1, 4]. Structure ordering has been detected in some of them.

One of the present authors [3] has shown that it is possible to form mixed crystals in the $Ga_2Se_3 - In_2Se_3$ system. The purpose of the present work was to investigate this system over a wide range of concentrations, to establish the limits of homogeneity of the ternary phase and the nature of interaction between the structural units used in substitution.

The initial components of the system — indium and gallium selenides — have different structures so that a study of this system is also of interest from the point of view of the influence of structure on the formation of solid solutions.

Ga_2Se_3 has the zinc blende structure, in which gallium atoms occupy, at random, two-thirds of the "cation sites" [5]. Our investigations have shown that the characteristic feature of the structure of Ga_2Se_3 appears in the powder diffraction patterns: the lines with odd indices are broad, and the lines $h + k + l = 4n$ remain sharp. All the broad lines are of the same width and shape if plotted in intensity coordinates: $-(\sin \delta)/\lambda$. A line shift is also observed so that the lattice period calculated from the positions of the sharp lines is different from the period calculated from the positions of the centers of the broad lines. The value of the shift and broadening of the lines with odd values of h, k, l, vary from sample to sample. These observations cannot be explained by lattice distortions or small dimensions of the crystals, or very small splitting of the lines due to distortion of the unit-cell shape; the cause should be sought in the distribution of gallium atoms in the tetrahedral spaces characteristic of Ga_2Se_3. This effect will be the subject of a later study.

In_2Se_3 is a representative compound of the $A_2^{III}B_3^{VI}$ type, which does not crystallize in the zinc blende structure. This compound has several modifications. One of them, stable above 200°C, has [6] the structure of a six-layer close packing of selenium atoms with indium atoms occupying, in an ordered manner, one-third of the tetrahedral spaces between selenium atoms. Thus, this structure is related to that of zinc blende. In the system considered, we investigated alloys along the tie-line $Ga_2Se_3 - In_2Se_3$ of the ternary system $In - Se - Ga$. All these alloys were studied by x-ray diffraction and their densities measured.

We prepared 12 alloys in all, including the initial components. They were synthesized by the usual melting of components.

X-ray diffraction patterns of annealed samples showed that all the alloys were close to the equilibrium state. Homogeneous ternary phases were formed with morphotropic transitions between structures over the whole range of concentrations. Alloys containing from 100 to 75% gallium selenide had the zinc blende structure. On increasing the In_2Se_3 content, the lattice parameter increased (Table 1). X-ray diffraction patterns of the alloy 75% $Ga_2Se_3 - 25\%$ In_2Se_3 exhibited, like Ga_2Se_3, broadening of the lines with odd indices.

TABLE 1

Composition	mol. %		Structure	Parameters in kXU		reduced		d_4^{20}	x-ray density g/cm³
	Ga₂Se₃	In₂Se₃		a	c	a	c		
Ga₂Se₃	100	—	sphalerite	5.41	—	3.89	9.40	—	—
7 Ga₂Se₃ · In₂Se₃ . . .	87,5	12,5	»	5.45	—	3.86	9.44	5.21	5.27
3 Ga₂Se₃ · In₂Se₃ . . .	75	25	»	5.49	—	3.88	9.51	5.24	5.38
7 Ga₂Se₃ · 3 In₂Se₃ . . .	70	30	wurtzite	3.93	6.42	3.93	9.63	5.21	5.17
Ga₂Se₃ · 2 In₂Se₃ . . .	66	34	»	3.95	6.46	3.95	9.70	5.16	5.12
29 Ga₂Se₃ · 21 In₂Se₃ . . .	58	42	»	3.95	6.46	3.95	9.70	5.20	5.22
Ga₂Se₃ · In₂Se₃ . . .	50	50	β—In₂Se₃	6.91	18.82	4.02	9.45	5.25	5.36
Ga₂Se₃ · 2 In₂Se₃ . . .	34	66	»	6.96	18.92	4.03	9.43	5.31	5.41
Ga₂Se₃ · 3 In₂Se₃ . . .	25	75	»	7.01	19.07	4.06	9.54	5.34	5.42
Ga₂Se₃ · 7 In₂Se₃ . . .	12,5	87,5	»	7.05	19.17	4.07	9.55	5.49	5.46
Ga₂Se₃ · 19 In₂Se₃ . . .	5	95	»	7.06	19.19	4.10	9.66	5.62	5.53
α—In₂Se₃	0	100	α—In₂Se₃	4.02	19.20	4.02	9.60	5.62	5.77
β—In₂Se₃	0	100	β—In₂Se₃	7.09	19.30	4.10	9.60	—	—

Alloys with from 70% Ga_2Se_3 to 58% In_2Se_3 had a structure close to that of wurtzite. The deviation of the alloy structure from the ideal wurtzite lattice was indicated by weak lines in the x-ray diffraction patterns, which could be indexed by trebling the unit-cell parameters listed for these alloys in Table 1. This suggests at least partial ordering.

Finally, the alloys containing from 50 to 5% In_2Se_3 had single-phase diffraction patterns. Since this region is close to In_2Se_3 and, as mentioned above, the structure of the β modification of In_2Se_3 is similar to the structure of Ga_2Se_3, we assumed that the solid solutions of this region had the structure of β-In_2Se_3 with a random replacement of some indium atoms by gallium atoms. All the x-ray diffraction lines could be indexed satisfactorily by assuming that the alloys had the same lattice as β-In_2Se_3, decreasing in size with the gradual replacement of indium with gallium. Calculation of the structure amplitudes, carried out even without substituting the numerical values of the atomic scattering factors, and comparison of the values obtained with the line intensity in the diffraction patterns, showed that the calculated and measured values of F^2 were different. However, a more detailed study showed that the differences between the calculated and observed intensities were due to small departures from the parameters listed by Semiletov [6] for β-In_2Se_3. Table 1 gives the values of the parameters calculated by us for all the investigated compositions. The same table lists, for comparison, the parameters of the high-temperature modification of β-In_2Se_3, taken from Semiletov's work. The parameters of the low-temperature modification of In_2Se_3 were calculated by us from the x-ray diffraction patterns of a single crystal. To make the comparison easier, Table 1 gives the "reduced parameters": "a," which is the diameter of a selenium atom, and "c," which is the thickness of three closely-packed layers. For alloys in region one (sphalerite structure), these reduced parameters are, respectively, half of the cube-face diagonal and the cube diagonal; for alloys in region two (wurtzite structure), the reduced parameters are: a and 1.5 c; for alloys in region three (β-In_2Se_3 structure) and for β-In_2Se_3 itself, they are: $a/\sqrt{3}$ and c/2; for the low-temperature phase, they are a and c/2. It is evident from Table 1 that the reduced parameters vary continuously with composition, having a small discontinuity on transition from alloys of one structure to those of another.

On the basis of the results obtained for the present system, we can draw the following conclusion. A series of solid solutions with morphotropic transition between structures is formed throughout the whole range of concentrations in the $Ga_2Se_3 - In_2Se_3$ system. When the indium selenide content is increased, the disordered structure of Ga_2Se_3 becomes partly ordered and similar to wurtzite, and it then transforms into the ordered β-In_2Se_3 structure. These results give us additional data on the possibility of the existence of conditions for morphotropic transitions in this group of substances. The system considered should be of great interest in respect to its physical properties since we can investigate the dependence of the properties on the type of structure and the degree of ordering.

The authors are grateful to N. A. Goryunova, Doctor of Chemical Sciences, for discussing the results and for her valuable advice.

LITERATURE CITED

1. V. S. Grigor'eva, Zh. Tekhn. Fiz. 28:1670 (1958).
2. J. C. Woolley and B. A. Smith, Proc. Phys. Soc. (London) 69:1339 (1958).
3. V. S. Grigor'eva, Dissertation, Leningrad (1959).
4. S. I. Radautsan and O. P. Derid, Izv. Akad. Nauk Moldav.SSR, No. 3(69):105 (1960).
5. H. Hahn and W. Klingler, Z. Anorg. Chem. B 250:97 (1949).
6. S. A. Semiletov, Fiz. Tverd. Tela 3:746 (1961).

SOME PROPERTIES OF TELLURIUM SINGLE CRYSTALS
CONTAINING SMALL AMOUNTS OF ANTIMONY AND ARSENIC

V. I. Veraksa, V. N. Lange, T. I. Lange, and A. R. Regel'

The possibility of changing considerably the properties of a semiconductor by introducing impurities is due not only to the appearance of impurity levels in the forbidden band but also to local structure changes due to the presence of foreign atoms in the lattice.

This conclusion follows, in particular, from those cases where the introduction of impurities of the same valence into a semiconductor sometimes does not greatly alter the semiconductors electrical properties, while in other cases, it produces a considerable nonmonotonic change in the carrier density and mobility. For example, tin introduced into germanium alters little its electrical properties up to concentrations of 10^{20} cm^{-3} [1], whereas relatively small amounts (0.05-0.10 at. %) of selenium or sulfur produce a sharp and nonmonotonic change in the electrical and other properties of tellurium [2-4].

Impurity atoms may enter a lattice at random or they may form some definite groups. The first case applies, of course, to very small amounts of impurities; it follows from the dominant entropy effects. However, when the concentration of the impurities is increased, impurity atom complexes may form. Such complexes were detected and investigated in the systems Cu−Au, Ag−Au, Al−Ag, and Al−Zn containing several tens of atomic per cent of one of the components [5-8].

It is interesting to investigate the transition from isolated impurity atoms to the formation of groupings.

There are reasons for assuming that the transition from a random distribution of impurity atoms to the formation of complexes should be accompanied by an "anomalous" nonclassical change in the properties, the changes being particularly marked in semiconductor alloys because of their characteristic directional covalent bonds. Goryunova [9] has shown that the more closely one approaches the covalent chemical binding, the more rigorous are the conditions for the formation of an "ideal" solid solution and the stronger is the dependence of the physical properties on structure changes in the alloy.

Our test materials were semiconductor alloys based on tellurium, whose characteristic chain structure [10] led us to expect particularly strong effects.

The first tests, the results of which were reported earlier [2-4], were carried out on tellurium−selenide and tellurium−sulfur systems. The nature of the dependences of the physical properties on composition was found by assuming that at low concentrations the atoms of selenium or sulfur enter the tellurium chains singly, but that on increasing the impurity content, the chains show, apart from the Te−Se−Te (or Te−S−Te) sequence, more complex sequences of the type Te−Se−Se−Se−Te (or, correspondingly, Te−S−S−S−Te). The results of an investigation of the electrical properties, density, and microhardness were in good agreement with this change of structure.

We have continued our study of the transition from one type of impurity distribution to another in the systems tellurium−antimony and tellurium−arsenic, which consist of elements belonging to different groups in the periodic system. In selecting these systems, we were influenced by the following considerations.

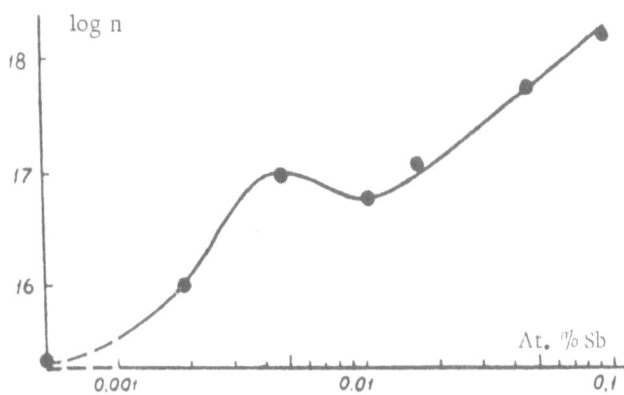

Fig. 1. Dependence of the carrier density on the sample's composition in the Te−Sb system. The curve is plotted for 90°K from the data reported in [11].

Having carefully analyzed the results of Fukuroi, Tanuma, and Tobisawa [11], who investigated the properties of tellurium single crystals doped with small amounts of antimony, we noted that at antimony concentrations of the order of 0.005 at. % the dependence of the carrier (hole) density on composition—plotted using the results of Fukuroi et al.—showed a considerable departure from monotonic behavior (Fig. 1). Fukuroi et al. [11] did not pay any attention to this point, regarding it as due to experimental scatter. However, we suggest that this deviation is not accidental but is of basic importance, indicating the complex nature of the formation of a solid solution of antimony in tellurium. Unfortunately, the region of antimony concentrations where this anomaly occurred was insufficiently investigated in the work of Kronmüller et al. [12]. Nevertheless, the results of Kronmüller et al. are not inconsistent with the possibility of a complex dependence of the carrier density on composition because one could plot either a monotonic or nonmonotonic curve through their data.

The stimulus to carry out the present investigation was provided by the results of Nasledov and Sokolov [13] and Andreev and Regel' [14] on iodine-doped selenium, which is a chemical and structural analog of tellurium. Both these studies demonstrated the anomalous change in the electrical properties (the electrical conductivity in weak and strong fields) at iodine concentrations of about 0.2-0.3 at. %. Andreev and Regel' suggested that the reason for these anomalies lies in the transition from an atomic to a molecular distribution of the impurities [14]. Similar anomalies were also observed by Abdullaev, Aliev, and Akhundova, who investigated the effect of bromine and thallium admixtures on the thermal conductivity of polycrystalline selenium and on its temperature dependence [15, 16]. The selenium-chlorine system also exhibits similar anomalies [17].

Apart from the work referred to earlier [11, 12], the effect of small amounts of antimony on the electrical properties of tellurium was also investigated by Cartwright and Haberfeld [18], who found that the addition of antimony to tellurium produces the same effects as the addition of group III elements to germanium or silicon.

Arsenic has the same structure as antimony [10], and has very similar chemical binding [19]. Therefore, we may expect that the system tellurium−arsenic will exhibit the same effects as the tellurium−antimony system. It is worth noting, too, that no anomalies were found in the only−according to our knowledge−work on the influence of arsenic on the properties of tellurium [12].

EXPERIMENTAL METHOD

Our samples were prepared from triply vacuum-distilled tellurium containing 10^{16} cm^{-3} active uncompensated impurities (as found from the value of the Hall coefficient in the impurity conduction region). The antimony was purified once by vacuum-distillation; the arsenic was of As$_{11}$ grade, obtained from the Rachinskii Mining and Metallurgical Combine.

We investigated alloys containing 0.000, 0.002, 0.005, 0.010, 0.020, and 0.050 at. % of the impurity. We used only seven compositions because our purpose was not to obtain the exact curves showing the dependence

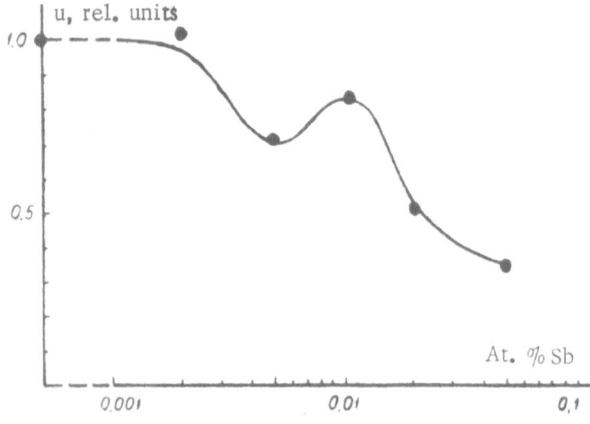

Fig. 2. Dependence of the carrier mobility on the composition of the Te−Sb system. The curve is plotted for 90°K, which is in the impurity-conduction region.

of a property on the composition−which might well be the subject of a special study−but only to check that these curves were not monotonic.

Single-crystal rods were grown by the Bridgman method from alloys prepared as above. We selected a growth regime which would ensure that the crystallographic axis of tellurium coincided with the sample axis. The electrical conductivity was measured by the usual dc potentiometric method. The Hall effect was investigated using alternating current and the apparatus described in [20]. The density was measured pycnometrically. Doubly distillled water, from which air was expelled by prolonged boiling, was used as the pycnometric liquid. We used a PMT-3 type instrument to measure the microhardness. The indenter was applied to the $(10\bar{1}0)$ face in such a way that one of the diagonals of the indentation made by a diamond pyramid was along the C axis. The hardness was estimated from the length of this diagonal. All measurements were carried out on several (not less than three) samples of the same composition; the results were then averaged out and the curves were plotted using only the average values. In all cases, the experimental errors were sufficiently small to avoid their masking the effects investigated here.

The results of the measurements of the electrical properties were analyzed using the formulas for an isotropic semiconductor with covalent binding.

RESULTS OF THE INVESTIGATION OF THE TELLURIUM−ANTIMONY SYSTEM

The temperature dependence of the electrical conductivity of all the samples of this system was of the typical semiconductor type. Increases in the antimony content led to a smooth rise in the electrical conductivity in the impurity conduction region. The value of the Hall coefficient also went up monotonically as the antimony concentration increased. However, the dependence of the carrier mobility on the alloy composition, plotted for 90°K (impurity conduction region) from the Hall effect and electrical-conductivity data, showed a clear minimum at 0.005 at. % Sb (Fig. 2).

The dependence of the density of the samples on composition (Fig. 3) also had a minimum, whose position was identical with the mobility minimum. The same can be said about the microhardness results, which are shown relative to tellurium in Fig. 4.

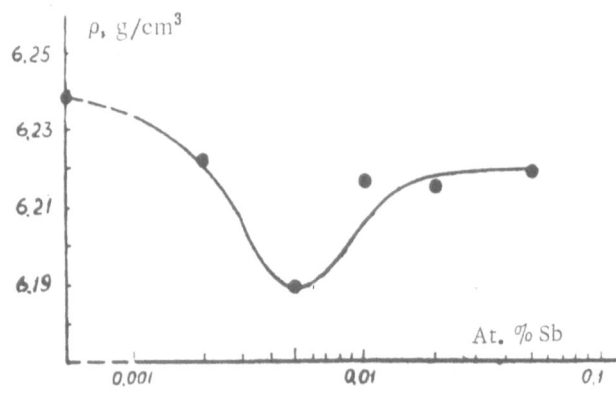

Fig. 3. Dependence, at room temperature, of the density of Te−Sb samples on their composition.

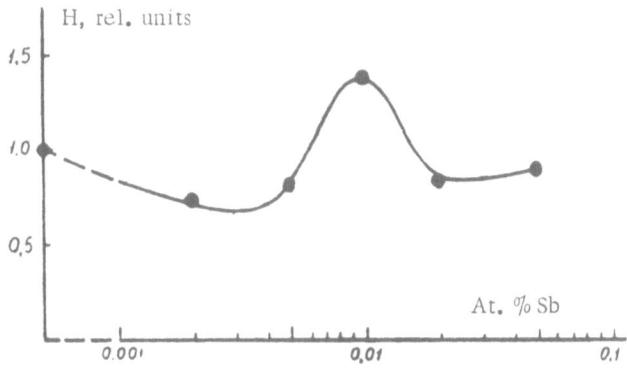

Fig. 4. Dependence of the microhardness on the composition in the Te−Sb system.

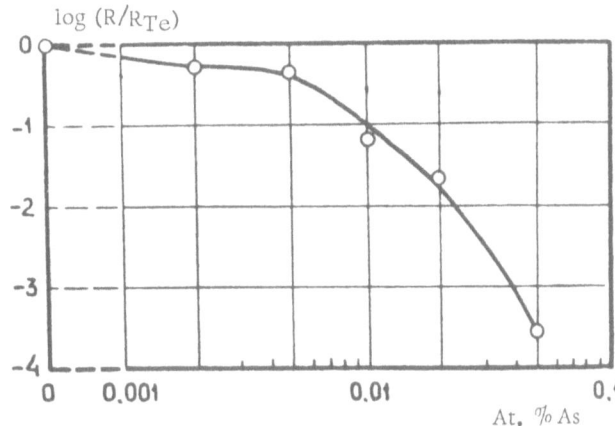

Fig. 5. Dependence of the Hall coefficient on the composition in the Te−As system at 90°K.

Later, a further series of tests was carried out in which samples were prepared from tellurium of a little higher purity. We obtained similar results for this series, except that the position of the minimum in the dependences of the density and of the carrier mobility (the hardness was not measured) was shifted a little toward lower antimony concentrations. This shift was not of basic importance and may have been due to several factors. First, the real and nominal antimony concentrations may be different and this difference may have had a different sign for the two series of samples. Secondly, the true position of the minimum in the dependence on the composition may have been somewhere between the antimony concentration values used in the two series of tests. The shift of the minimum could also have been affected by differences in the concentration and nature of impurities in the initial tellurium. Finally, the two series of samples may have had a different number of defects because of uncontrolled details in the technology of the samples' preparation.

Figure 2 shows that the interaction of small amounts of antimony reduces the hole mobility in contrast to selenium and sulfur [2-4], which increase this mobility. This must be due to the strong scattering of carriers on ionized single atoms of antimony. The effect of chain "closure," suggested as an explanation of the rise of the hole mobility in tellurium on the introduction of selenium or sulfur [4], does not take place in the case of antimony because of the method by which impurities are introduced into the tellurium lattice, a matter which will be discussed in detail below.

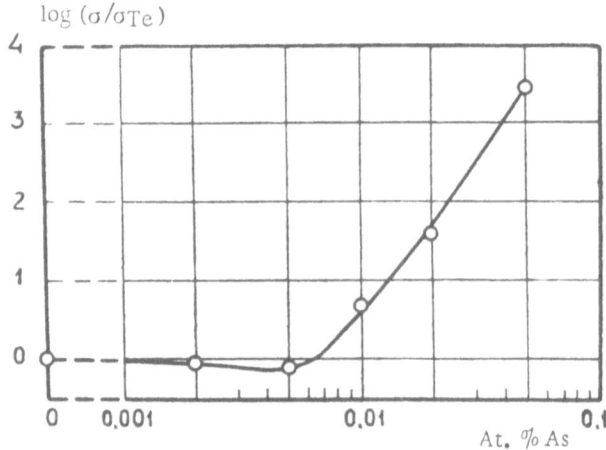

Fig. 6. Electrical conductivity of the Te−As system.
The curve is plotted for 90°K.

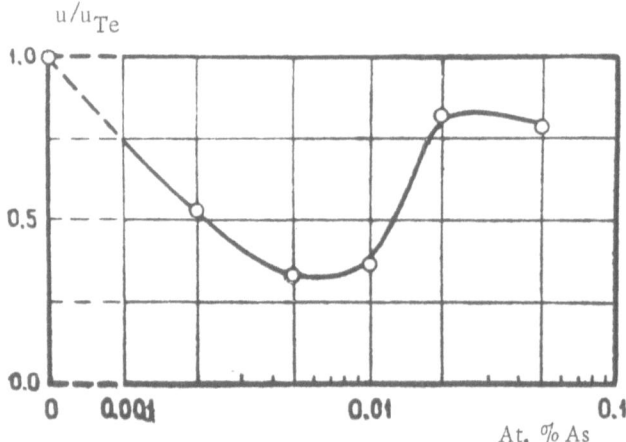

Fig. 7. Hole mobility in the Te−As system, calculated
from the data on the Hall effect and electrical conductivity at 90°K.

RESULTS OF THE INVESTIGATION OF THE TELLURIUM−ARSENIC SYSTEM

The results obtained for the tellurium−arsenic system are shown in Figs. 5-8. Figure 5 gives in relative units the dependence of the Hall coefficient on composition. It is evident from this that the number of carriers depends little on the arsenic content up to 0.005 at. %. However, above this concentration, the carrier density begins to rise and at 0.050 at. % increases by over three orders of magnitude. The sign of the Hall coefficient in samples of all compositions corresponds to p-type conduction, as in the tellurium−antimony system.

Although the Hall coefficient varies monotonically with composition, the dependence of the electrical conductivity is somewhat more complex (Fig. 6). First, the electrical conductivity falls, reaching a minimum about 20% lower than that in pure tellurium (this occurs at about 0.005 at. % As) and it then increases rapidly by more than three orders of magnitude.

The deviation from the monotonic variations is even more marked in the curve shown in Fig. 7. The carrier mobility decreases by 70% when 0.005 at. % As is introduced; it then increases, almost approaching its value in tellurium and, finally, above 0.020 at. % As, begins to decrease again. If arsenic forms the usual substitutional or interstitial solid solutions in tellurium, we would expect, in accordance with Vegard's law, the

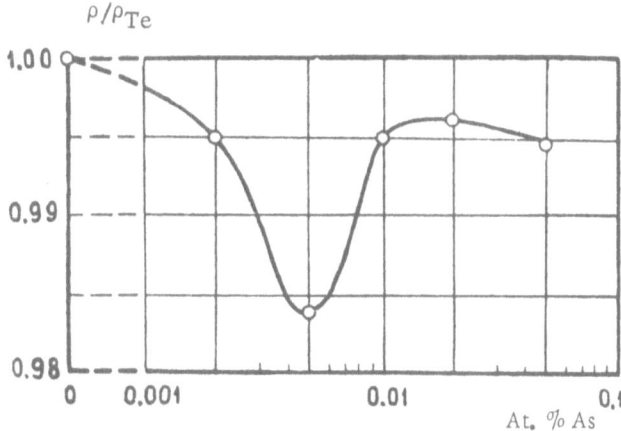

Fig. 8. Density in the Te−As system. The data refer to
room temperature (as Fig. 3).

composition dependence of the density to be monotonic, and to be little different from the density of pure tellurium in the impurity concentration range 0-0.050 at. %. In fact, Fig. 8 shows that even 0.005 at. % arsenic reduces the density by 2%. When the arsenic content is increased, the density of the alloys becomes almost equal to the density of tellurium.

Thus, the dependence of the various properties on the composition of the tellurium−arsenic system is basically the same as in the tellurium−antimony system.

DISCUSSION OF THE RESULTS

We shall assume that structural changes in the systems tellurium−antimony and tellurium−arsenic are basically the same as in the systems tellurium−sulfur and tellurium−selenium, i. e.,when the impurity concentration is increased the number of defects first increases and then begins to decrease when complexes start forming. However, the nature of defects due to single impurity atoms and their groups may differ considerably for different impurities. In the tellurium−selenium and tellurium−sulfur systems, the experimental results were in best agreement with the assumption that while single atoms of selenium or sulfur, entering tellurium chains, bend and then break these chains, complexes consisting of three impurity atoms in a sequence reestablish the original structure and properties of tellurium.

Due to the great differences between the structures (chain structure of tellurium and layered structure of antimony and arsenic [10]), it is natural to assume that the nature of the defects appearing in the tellurium lattice on the introduction of antimony and arsenic should be different from the case when the impurity is selenium, which has the same structure as tellurium, or sulfur, which has a similar structure. It is quite likely that single atoms of antimony or arsenic, as well as complexes of such atoms, do not join tellurium chains but become located between such chains. This is confirmed by studies of the microhardness of the cleavage planes of single-crystal samples in the tellurium−selenium and tellurium−antimony systems.

A continuous increase in the ratio of the long-to-short diagonal in the indentation made by the diamond pyramid observed in the tellurium−selenium system on increase in the selenium content (Table 1) indicates a relative increase in the strength of the internal bonds in the chains at the expense of a weakening of the bonds between chains. This may be explained by the stronger saturation of the bonds between atoms in the chains in the presence of selenium, because the Te−Se bond is somewhat more stable than the Te−Te bond.

On the other hand, antimony produces a monotonic reduction in both indentation diagonals, and their ratio remains, within experimental error, equal to the ratio of the diagonals for pure tellurium (Table 2).

TABLE 1. Change in the Ratio of the Diagonals of a Diamond Pyramid
Indentation in the Measurement of the Microhardness in the
Tellurium—Selenium System*

At. % Se	0.00	0.10	0.20	0.50	1.00
Ratio of diagonals	1.00	1.04	1.08	1.42	1.47

*The ratio of diagonals in pure tellurium was taken to be unity.

TABLE 2. Change in the Ratio of the Diagonals of a Diamond Pyramid
Indentation in the Measurement of the Microhardness in the
Tellurium—Antimony System*

At. % Sb	0.000	0.002	0.005	0.010	0.020	0.050
Ratio of diagonals	1.00	1.00	0.98	0.99	1.00	0.98

*The ratio of diagonals in pure tellurium was taken to be unity.

We shall assume that single atoms of antimony and arsenic become located mainly between tellurium chains, pushing them apart and bending them. In this way, "voids" are formed which reduce the hardness and density. The greater number of lattice defects reduces the carrier mobility. Since arsenic and antimony atoms are easily ionized, the carrier density increases.

Two-dimensional layer complexes of impurity atoms are formed along planes parallel to the C axis when the impurity concentration is increased. Such layers distort the tellurium structure somewhat and, therefore, the values of the density, microhardness, and carrier mobility increase. The carrier density may also be reduced due to formation of impurity complexes. This would explain the anomalies of the carrier-density dependence on composition, which are observed (Fig. 1) if the initial tellurium is sufficiently pure.

Attention should be drawn to the fact that the introduction of antimony and arsenic impurities in amounts of 0.005 at. % reduces the sample density by 1-2%. Comparison of these two values leads to the conclusion that each impurity atom in the tellurium lattice forms a void of the order of 100-200 atomic volumes.

Summarizing, we can say that it is reliably established that some physical properties of tellurium vary nonmonotonically at low concentrations of antimony and arsenic, as indicated by studies of individual properties of the tellurium—antimony and tellurium—arsenic systems as well as by all the data taken together, which show good correlation. This is also confirmed by similar effects observed earlier for several systems based on tellurium and selenium [2-4, 13-17, 21].

However, the available data are insufficient to assert that the proposed mechanism of the behavior of antimony and arsenic atoms in tellurium is fully proved. Further studies are needed to obtain complete information on the structure and defect properties.

In conclusion, the authors thank A. V. Novoselova, Corresponding Member of the USSR Academy of Sciences, for discussing the results of the present work.

LITERATURE CITED

1. F. Herman, M. Gliksman, and R. H. Parmenter, Progr. in Semicond. 2:1 (1957).
2. V. N. Lange and A. R. Regel', Fiz. Tverd. Tela 1:559 (1959).
3. V. N. Lange and A. R. Regel', Fiz. Tverd. Tela 1:562 (1959).
4. V. N. Lange and A. R. Regel', Fiz. Tverd. Tela 2:2439 (1960).
5. J. M. Cowley, J. Appl. Phys. 21:24 (1950).
6. N. Norman and B. E. Warren, J. Appl. Phys. 22:483 (1951).
7. C. B. Walker, J. Blin, and A. Quinier, Compt. Rend. 235:254 (1952).
8. P. S. Rudman, P. A. Flinn, and B. L. Averbach, J. Appl. Phys. 24:365 (1953).
9. N. A. Goryunova, Dissertation for the degree of Doctor of Chemical Sciences, IONKh (1958).
10. B. F. Ormont, Structure of Inorganic Substances, GITTL (1958).
11. T. Fukuroi, S. Tanuma, and S. Tobisawa, Sci. Rept. Res. Inst. Tohoku Univ. Ser. A 4:283 (1952).
12. H. Kronmüller, J. Jaumann, and K. Seiler, Z. Naturforsch. 11a(3):243 (1956).
13. D. N. Nasledov and B. V. Sokolov, Zh. Tekhn. Fiz. 28(4) (1958).
14. A. A. Andreev and A. R. Regel', Fiz. Tverd. Tela 2:2770 (1960).
15. M. I. Aliev and G. B. Abdullaev, Fiz. Tverd. Tela 1:1296 (1959).
16. G. B. Abdullaev, M. I. Aliev, and S. A. Akhundova, Fiz. Tverd. Tela 3:326 (1961).
17. G. M. Aliev and G. B. Abdullaev, Dokl. Akad. Nauk SSSR 98(4):557 (1954).
18. C. H. Cartwright and M. Haberfeld, Nature 134(3382):287 (1934).
19. B. V. Nekrasov, General Chemistry Course, Goskhimizdat, Moscow (1952).
20. V. N. Lange, Sb. Nauchn. Rabot Barnaul'sk. Gos. Ped. Inst. No. 3:357 (1958).
21. Yin Shih-tuan and A. R. Regel', Fiz. Tverd. Tela 3:1688 (1961).

ANISOTROPY OF THE GALVANOMAGNETIC PROPERTIES
OF BISMUTH SINGLE CRYSTALS
AND THEIR ALLOYS WITH TELLURIUM

D. V. Gitsu and G. A. Ivanov

INTRODUCTION

The present paper reports the results of a study of the influence of tellurium on the anisotropy of the galvanomagnetic properties of bismuth single crystals. Some of these results were published earlier [1, 2].

The study of the anisotropy and its changes due to the presence of small amounts of foreign atoms are of interest for the following reasons.

1. A careful study of the dependence of the physical properties on the orientation of the crystallographic axes can give us more extensive and reliable information on the processes occuring in crystalline solids than can be obtained from studies of quasi-isotropic polycrystalline media. The elucidation of these processes helps in the further development and refinement of the theory which, in the majority of cases, gives at present only a qualitative interpretation of the observed effects.

2. Since the anisotropy of the electrical properties is related to the structure of the energy spectrum of electrons in a crystal, this anisotropy may be described theoretically only if an approximate model of the energy spectrum is correctly selected. If we neglect the deformation of the lattice due to the introduction of impurities (at low concentrations), the nature of the change in the anisotropy is governed primarily by the energy band population and the behavior of the density of states along various crystallographic directions. Consequently, a change in the anisotropy due to impurities may yield very useful information on the band structure of a crystal in the regions away from the extrema.

The method of growing single crystals of bismuth and its alloys and the methods of investigating them have been described in the published work [1, 2]. We shall only mention here that we investigated single-crystal samples of the following three orientations: A) the trigonal axis parallel to the sample axis; B) one of the binary axes parallel to the sample axis; C) the trigonal axis and one of the binary axes, both perpendicular to the sample axis.

EXPERIMENTAL RESULTS

To make the comparison with theory easier, the experimental results are presented as follows:

a) Figures 1-3 show the rotation diagrams $R(\Theta)$ and $\Delta\rho/\rho$ (Θ) for pure bismuth in fields of 5000, 12,200, and 18,000 Oe; anisotropy appears very clearly in these samples, so that it is easy to check whether it agrees with the theory;

b) the same diagrams are given for bismuth-tellurium alloys at H = 5000 Oe in Figs. 4-9; we can deduce from these diagrams the influence of the tellurium impurity not only qualitatively, but we can also find the order of magnitude of the components of the tensors $R_{ij,k}$ and $\rho_{ij,kl}$ and thus determine quantitatively the change in the galvanomagnetic effects.

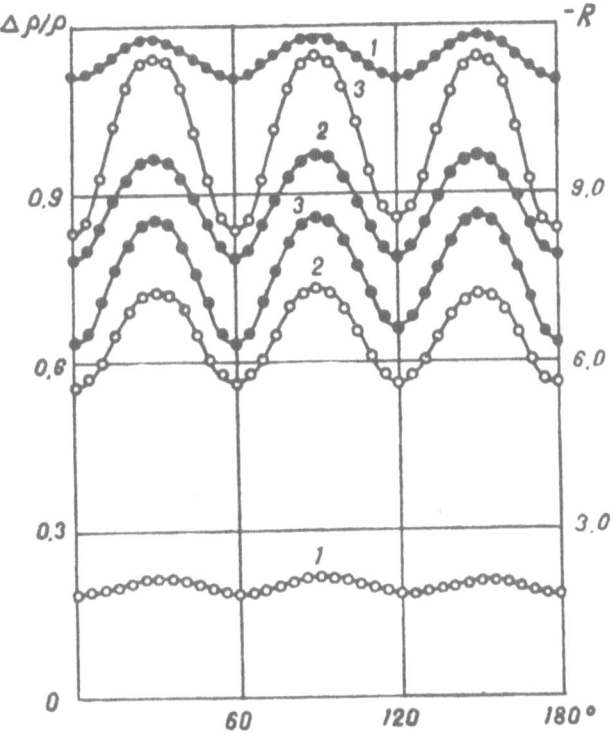

Fig. 1. Angular dependence ("rotation diagrams") of the Hall coefficient R(Θ) and of the magnetoresistance Δρ/ρ(Θ) for type A samples. Here and later, the black dots denote the Hall coefficient, and the open circles denote the magnetoresistance. 1) H = 5000 Oe; 2) H = 12,200 Oe; 3) H = 18,000 Oe.

We shall now consider separately the curves for samples of each type.

1. When samples of type A are rotated, the magnetic field remains perpendicular to the trigonal axis, along which the current flows, but the mutual positions of the binary axes and the field are altered. When a sample is rotated by 180° the rotation diagrams R(Θ) and Δρ/ρ (Θ) pass three times through a maximum and minimum of equal amplitude at equal angular intervals (Figs. 1, 4, and 5). The minima are obtained when one of the binary axes is parallel to the magnetic field, the maxima when one of these axes is perpendicular to the field. The symmetry of R(Θ) and Δρ/ρ(Θ) is the same. The Hall coefficient is negative for all directions and its absolute value decreases with increasing magnetic field intensity.

The absolute magnitude of the galvanomagnetic effects drop sharply on increase in the tellurium concentration. The amplitudes of the variation in R(Θ) and Δρ/ρ(Θ) also decrease, and in samples containing more than 0.3% Te, the symmetry is practically circular, even in a field H = 18,000 Oe [2].

2. In crystals of type B, the axis about which the sample is rotated is parallel to one of the binary axes, along which the current flows. The magnetic field is perpendicular to the trigonal axis in the Θ = 90° position. The symmetries of the R(Θ) and Δρ/ρ(Θ) are now different (Figs. 2, 6, 7).

The Hall coefficient passes through a sharp minimum and maximum when a sample is rotated by 180°. At the minimum (Θ ≈ 5–10°) R(Θ) is positive, but it is negative in the maximum position; Δρ/ρ (Θ) has a minimum shifted by about 20–25° from Θ = 0°. The form of the Δρ/ρ(Θ) curves of pure bismuth depends strongly on H. When the magnetic field intensity is increased, the minimum becomes considerably deeper and the maximum is replaced by a plateau. The maximum itself shifts considerably to the right and becomes stronger. In weak magnetic fields, the symmetry of these curves is not greatly affected by the addition of tellurium as an impurity, but the absolute magnitudes of the galvanomagnetic effects and their dependence on H decrease very rapidly.

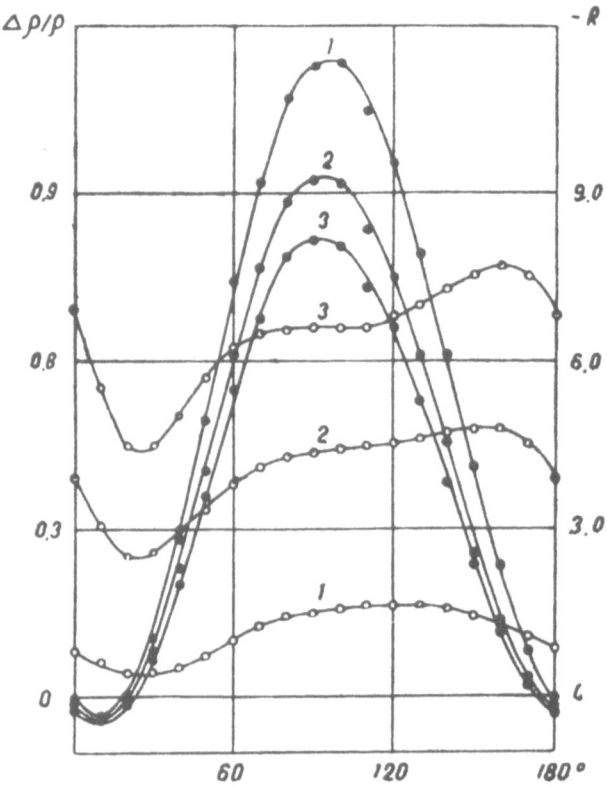

Fig. 2. R(Θ) and Δρ/ρ (Θ) diagrams for B-type samples.

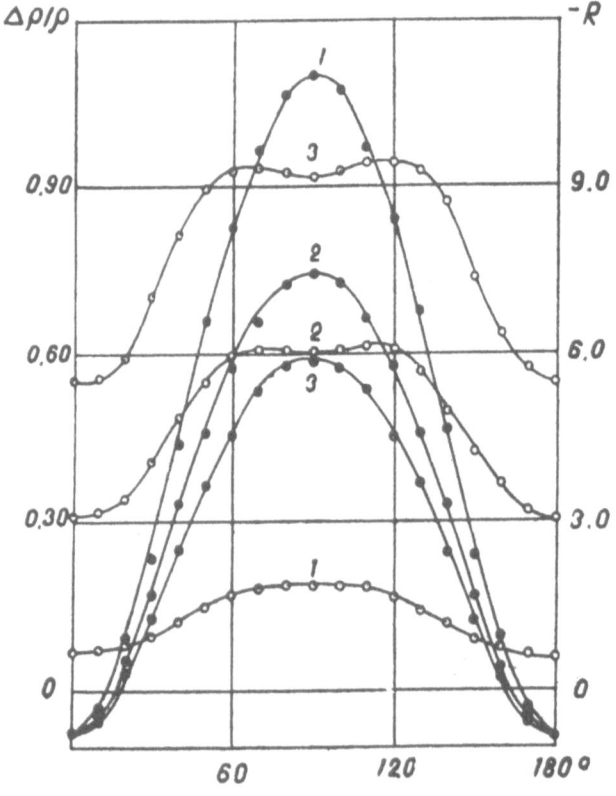

Fig. 3. R(Θ) and Δρ/ρ(Θ) diagrams for C-type samples.

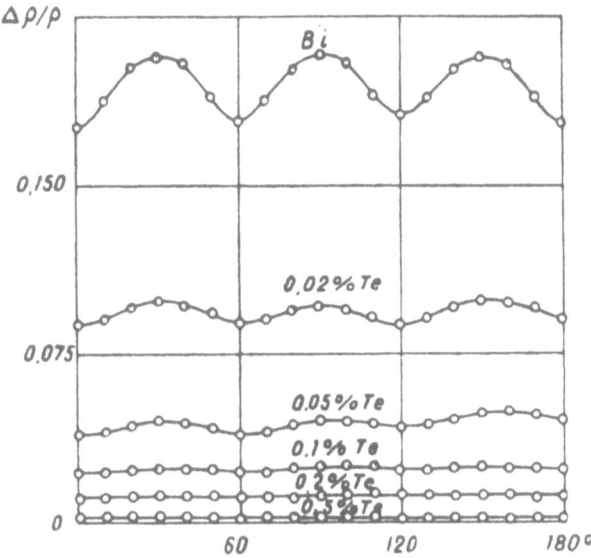

Fig. 4. $\Delta\rho/\rho(\Theta)$ diagrams for A-type samples of Bi—Te alloys in a field H = 5000 Oe.

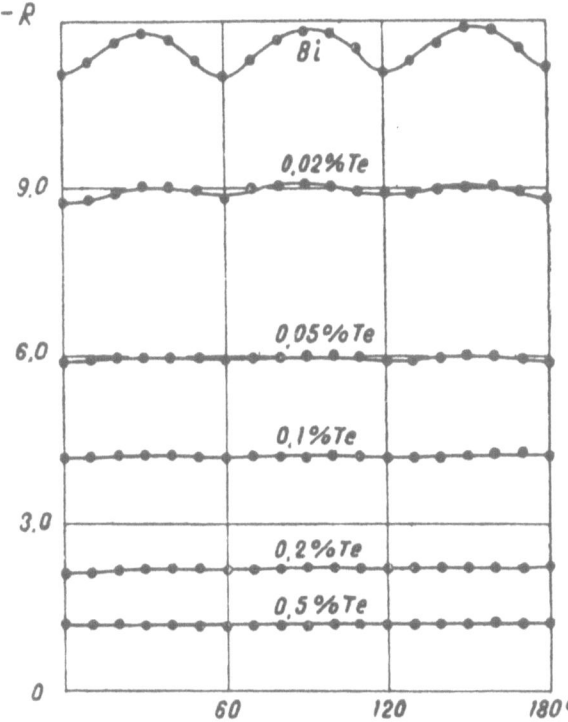

Fig. 5. R(Θ) diagrams for A-type samples of Bi—Te alloys in a field H = 5000 Oe.

3. In type C crystals, the rotation axis and the direction of the current are perpendicular to the trigonal axis and to one of the binary axes. The positions of the trigonal and all three binary axes change during rotation. At $\Theta = 0°$, the trigonal axis is parallel to H; at $\Theta = 90°$ it is perpendicular to this field. In the former case, we observe a minimum in the R(Θ) and $\Delta\rho/\rho$ (Θ) curves; in the latter case we have a maximum (Figs. 3, 8, 9). As in the case of type crystals, the symmetry of the two curves is different, particularly in strong fields.

For given H, the value of R depends first on the trigonal axis orientation. This orientation also has a strong influence of the magnitude of the magnetoresistance in weak magnetic fields. However, if strong fields are applied to pure bismuth, the maximum is replaced by a second minimum, associated with the orientation of the binary axes.

The addition of tellurium alters the diagrams for C-type crystals in the same way as the diagrams for B-type crystals.

PHENOMENOLOGICAL THEORY
OF GALVANOMAGNETIC EFFECTS
IN CRYSTALS OF THE D_{3d} CLASS

The relation between the anisotropy and crystal lattice symmetry is easiest to find using the phenomenological theory. In isotropic media in the absence of a temperature gradient, the density of a current $\vec{I}$, due to a field $\vec{E}$, is given by the generalized Ohm law:

$$E_i = P_{ij}(\vec{H})I_j, \tag{1}$$

where E_i is the electric field component in the i-th direction, and P_{ij} is the resistance tensor which is a function of the magnetic field H. According to Onsager's principle,

$$P_{ij}(\vec{H}) = P_{ji}(-\vec{H}). \tag{2}$$

Separating P_{ij} (H) into the symmetric and antisymmetric tensor components, we obtain

$$E_i = \rho_{ij}(\vec{H})\,I_j + R_{ij}(\vec{H})I_j, \tag{3}$$

where $\rho_{ij}(\vec{H})$ is an even, and $R_{ij}(\vec{H})$ is an odd function of $\vec{H}$. The former may be regarded as the generalized magnetoresistance, and the latter as the generalized Hall effect.

Equation (3) is the initial equation of the phenomenological theory of galvanomagnetic effects [3, 4].

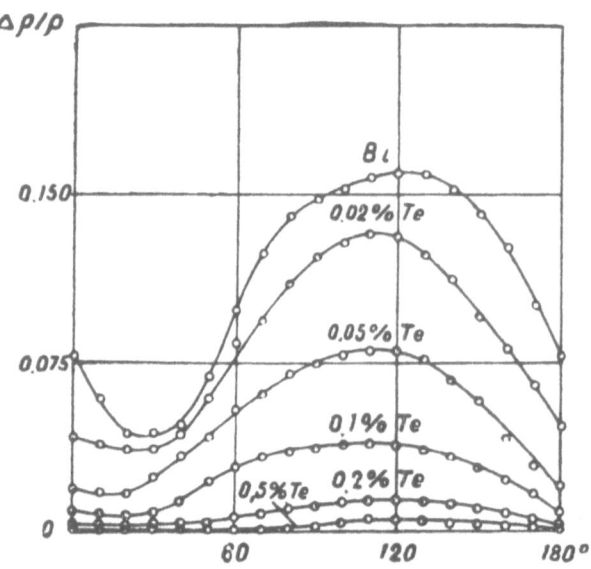

Fig. 6. $\Delta\rho/\rho(\Theta)$ diagrams for B-type samples of Bi−Te
alloys in a field H = 5000 Oe.

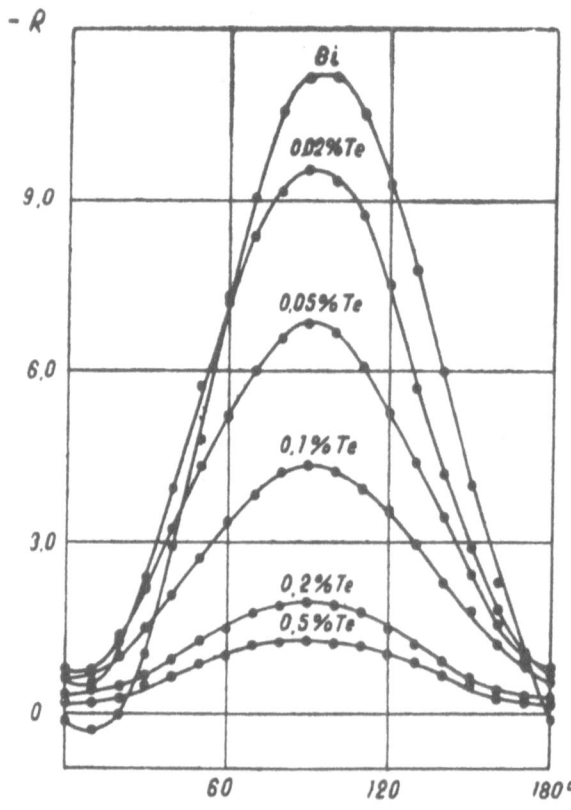

Fig. 7. R(Θ) diagrams for B-type samples of Bi−Te
alloys in a field H = 5000 Oe.

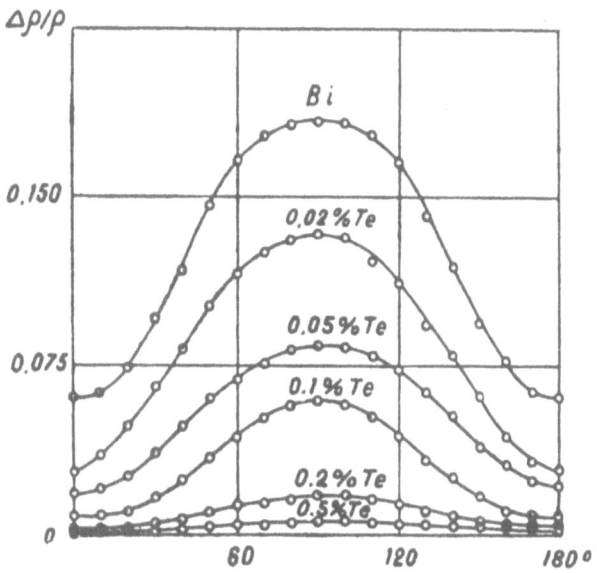

Fig. 8. $\Delta\rho/\rho(\Theta)$. C-type samples of Bi−Te alloys,
H = 5000 Oe.

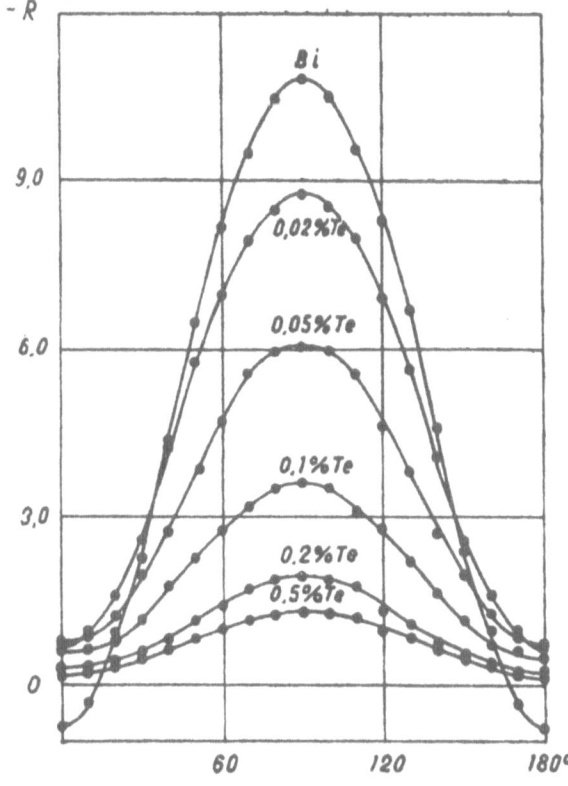

Fig. 9. R(Θ). C-type samples of Bi−Te alloys,
H = 5000 Oe.

In weak magnetic fields, $\rho_{ij}(\vec{H})$ and $R_{ij}(\vec{H})$ can be expressed as series in powers of H. Such expansion gives:

$$E_i = \rho_{ij}I_j + \rho_{ij,\,kl}I_jH_kH_l + \rho_{ij,\,klmn}I_jH_kH_lH_mH_n + \ldots + R_{ij,\,k}I_jH_k + R_{ij,\,klm}I_jH_kH_lH_m + \cdots \quad (4)$$

The tensors occurring in equation (4) obey the following self-evident relationships:

$$
\begin{aligned}
&\rho_{ij} = \rho_{ji},\\
&\rho_{ij,\,kl} = \rho_{ji,\,kl} = \rho_{ij,\,lk},\ldots\\
&\rho_{ij,\,klmn} = \rho_{ji,\,klmn} = \rho_{ij,\,lmnk},\ldots\\
&R_{ij,\,k} = -R_{ji,\,k},\\
&R_{ij,\,klm} = -R_{ji,\,klm} = R_{ij,\,lmk}\ldots
\end{aligned}
\quad (5)
$$

Using symmetry elements of the D_{3d} class, we can show that the number of independent tensor components for crystals isomorphous with bismuth are, respectively: two for ρ_{ij} and $R_{ij,\,k}$, six for $R_{ij,\,klm}$, eight for $\rho_{ij,\,kl}$, and eighteen for $\rho_{ij,\,klmn}$. The total numbers of independent components of $R_{ij,\,klmno}$ and $\rho_{ij,\,klmnop}$ have not been calculated.

If we select the coordinate system so that the trigonal axis is along the z axis and one of the binary axes along the x axis, then these tensors become:

$$
\rho_{ij} = \begin{Bmatrix} \rho_{11} & 0 & 0 \\ 0 & \rho_{11} & 0 \\ 0 & 0 & \rho_{33} \end{Bmatrix}
\quad (6)
$$

$$
R_{ij,\,k} = \begin{Bmatrix} R_{12,3} & 0 & 0 \\ 0 & R_{23,1} & 0 \\ 0 & 0 & R_{23,1} \end{Bmatrix}
\quad (7)
$$

$$
\rho_{ij,\,kl} = \begin{Bmatrix}
\rho_{11,11} & \rho_{11,22} & \rho_{11,33} & \rho_{11,22} & 0 & 0 \\
\rho_{11,22} & \rho_{11,11} & \rho_{11,33} & -\rho_{11,23} & 0 & 0 \\
\rho_{33,11} & \rho_{33,11} & \rho_{33,32} & 0 & 0 & 0 \\
\rho_{23,11} & -\rho_{23,11} & 0 & \rho_{23,23} & 0 & 0 \\
0 & 0 & 0 & 0 & \rho_{23,23} & \rho_{23,11} \\
0 & 0 & 0 & 0 & \rho_{11,23} & \frac{1}{2}(\rho_{11,11}-\rho_{11,22})
\end{Bmatrix}
\quad (8)
$$

$$R_{ij,\,klm} =$$

$$
\begin{Bmatrix}
0 & R_{12,112} & R_{12,113} & 0 & 0 & 0 & -R_{12,112} & R_{12,113} & 0 & R_{12,333} \\
0 & -\frac{1}{3}R_{23,111} & -R_{23,123} & 0 & 0 & 0 & -R_{23,111} & R_{23,123} & -R_{23,133} & 0 \\
R_{23,111} & 0 & 0 & \frac{1}{3}R_{23,111} & R_{23,123} & R_{23,133} & 0 & 0 & 0 & 0
\end{Bmatrix}
\quad (9)
$$

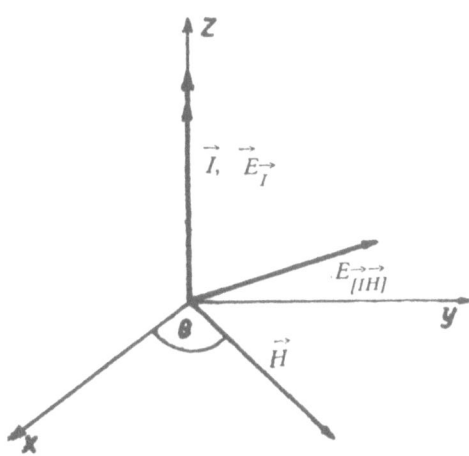

Fig. 10. Type A sample. The orientation of the current $\vec{I}$, magnetic field $\vec{H}$, and electric field components $E_{\vec{I}}$ and $E_{[\vec{I}\vec{H}]}$ with respect to the crystallographic axes.

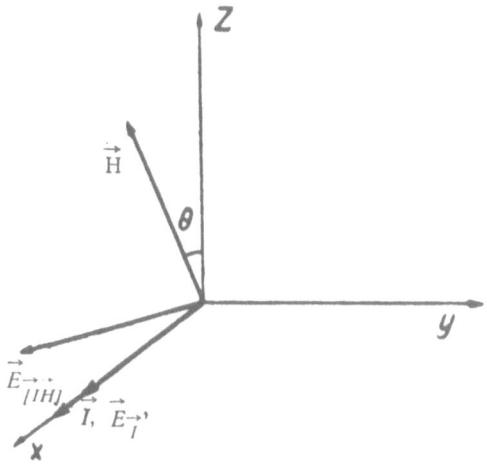

Fig. 11. Type B sample. The orientation of the current $\vec{I}$, magentic field $\vec{H}$, and electric field components $E_{\vec{I}}$ and $E_{[\vec{I}\vec{H}]}$ with respect to the crystallographic axes.

The nonzero components of the tensor $\rho_{ij,\ klmn}$ are listed in Table 1, where the grouping in the columns is in accordance with the indices ij, and that in the rows in accordance with the indices klmn.

Substituting the tensor components into Eq. (4), we obtain general expressions for the electric field E_1, E_2, and E_3 in terms of the current-density $\vec{I}$ and magnetic-field $\vec{H}$ components.

The directions of the current $\vec{I}$ and of the magnetic field $\vec{H}$ are known from investigations of the galvanomagentic effects. Usually, the following variants are employed:

1. $\vec{I} \perp \vec{H}$. : a) the electric field $E_{\vec{I}}$ is measured along the current, and b) the electric field is measured along the $[\vec{I}\vec{H}]$ direction.

2. $\vec{I} \parallel \vec{H}$: $E_{\vec{I}}$ is measured.

Here, we are interested only in the first variant. We shall now consider each type of sample separately.

Type A Samples

Figure 10 shows that in this case:

$$
\begin{array}{lll}
I_1 = 0, & H = H \cos \Theta, & \\
I_2 = 0, & H = H \sin \Theta, & \\
I_3 = I, & H = 0, &
\end{array}
\tag{10}
$$

For the electric field component, we have:

$$
E_{\vec{I}} = E_3,
\tag{11}
$$

$$
E_{[\vec{I}\vec{H}]} = -E_1 \sin\Theta + E_2 \cos\Theta.
\tag{12}
$$

If we substitute the values of E_1, E_2, and E_3 into Eqs. (11) and (12) and take only the terms as high as H^4, we find that no angular dependence of the Hall coefficient or the magnetoresistance is obtained. This is realized in practice only in very weak magnetic fields and in alloys containing large amounts of tellurium. To obtain the periodic variation in the galvanomagnetic effects in strong fields it is necessary to include at least the terms

TABLE 1

$\frac{ij}{klmn}$	11	22	33	23	13	12
1111	$\rho_{11,1111}$	$\rho_{22,1111}$	$\rho_{33,1111}$	$(\rho_{23,1122}+2\rho_{13,1222})$	0	0
2222	$\rho_{11,2222}$	$(\rho_{11,1111}+\rho_{22,1111}-\rho_{11,2222})$	$\rho_{33,1111}$	$-(3\rho_{23,1122}+2\rho_{13,1222})$	0	0
3333	$\rho_{11,3333}$	$\frac{1}{6}(\rho_{11,1111}-2\rho_{22,1111}\,3\rho_{11,2222})$	$\rho_{33,3333}$	0	0	0
1122	$\frac{1}{6}(\rho_{11,1111}+4\rho_{22,1111}-3\rho_{11,2222})$	0	$\frac{1}{3}\rho_{33,1111}$	$\rho_{23,1122}$	0	0
1133	$\rho_{11,1133}$	$\frac{1}{2}(\rho_{11,1123}-\rho_{11,2223})$	$\rho_{33,1133}$	$\rho_{23,1133}$	$\rho_{13,1123}$	0
2233	$\rho_{11,2233}$	0	$\rho_{33,1133}$	$-\rho_{13,1233}$	$(2\rho_{23,1122}+\rho_{13,1222})$	0
1123	$\rho_{11,1123}$	$-\rho_{11,2223}$	$\rho_{3,1123}$	$\rho_{23,1123}$	0	0
1113	0	0	0	0	$\frac{1}{3}\rho_{13,1222}$	$\frac{3}{4}(\rho_{11,1132}+\rho_{11,2223})$
1112	0	0	0	0	$\rho_{13,1222}$	$\frac{1}{4}(\rho_{11,1111}+2\rho_{22,1111}-3\rho_{11,2222})$
2223	$\rho_{11,2233}$	$-\frac{1}{2}(3\rho_{11,1123}+\rho_{11,2223})$	$-\rho_{33,1123}$	$\rho_{23,1123}$	0	0
2231	0	0	0	0	$\rho_{13,1333}$	$\frac{1}{4}(\rho_{11,1123}+\rho_{11,2223})$
2212	0	0	0	0	0	$\frac{1}{4}(\rho_{11,1111}-\rho_{22,1111}+\rho_{11,2222})$
3323	$\rho_{11,3333}$	$-\rho_{11,2333}$	0	$\rho_{13,1333}$	0	0
3331	0	0	0	0	0	$\rho_{11,2333}$
3312	0	0	0	0	$\rho_{13,1233}$	$\frac{1}{2}(\rho_{22,2233}-\rho_{11,2233})$

TABLE 2

Alloy composition	$\rho_{11} \times 10^4$, ohm-cm	$\rho_{33} \times 10^4$, ohm-cm	$R_{23,1}$, cgs emu	$R_{12,3}$, cgs emu	$\rho_{11,11} \times 10^{12}$, ohm-cm/Oe^2	$\rho_{11,22} \times 10^{12}$, ohm-cm/Oe^2	$\rho_{11,33} \times 10^{12}$, ohm-cm/Oe^2	$\rho_{33,11} \times 10^{12}$, ohm-cm/Oe^2	$\rho_{33,33} \times 10^{12}$, bhm-cm/Oe^2
Bi	1.05	1.37	—14.0	+0.8	0.24	1.13	0.25	1.70	0.07
Bi—0.02% Te	1.02	1.14	—10.3	—0.7	0.15	0.77	0.20	1.00	0.03
Bi—0.05% Te	0.94	0.97	— 6.6	—0.7	0.09	0.30	0.08	0.23	0.02
Bi—0. 1% Te	0.88	0.94	— 4.5	—0.6	0.04	0.13	0.04	0.83	0.01
Bi—0. 2% Te	0.84	0.88	— 2.2	—0.3	0.02	0.063	0.01	0.042	—
Bi—0. 3% Te	0.85	0.88	— 1.5	—0.2	—	0.023	0.005	0.014	—
Bi—0. 4% Te	0.86	0.98	— 1.5	—0.2	—	0.019	0.004	0.012	—
Bi—0. 5% Te	0.87	1.00	— 1.4	—0.2	—	0.015	0.004	0.010	—

containing H^5 and H^6. It follows from Eqs. (10)-(12) that, in type A crystals, it is sufficient to include only some components of the $R_{13,klmno}$, $R_{23,klmno}$, and $\rho_{33,klmno}$ types. For the nonzero components, we obtain:

$$R_{13,11112} = -R_{23,11122}, \quad R_{13,2222} = (3R_{13,11112} + 2R_{13,11222}),$$

$$R_{13,11222} = -R_{23,1222}, \quad R_{23,1111} = -(2R_{13,11112} + 3R_{13,11222}). \tag{13}$$

$$\rho_{33,111122} = \frac{1}{5}(-2\rho_{33,111111} + 3\rho_{33,222222}),$$

$$\rho_{33,112222} = \frac{1}{5}(3\rho_{33,111111} - 2\rho_{33,222222}). \tag{14}$$

Other components of the same type, containing the index 3 after the comma, are not listed because $H_3 = 0$. Including the terms with H^5 and H^6, we now have

$$E\vec{I} = \rho_{33}I + \rho_{33,11}IH^2 + \rho_{33,1111}IH^4 + (\rho_{33,111111}\cos^2 3\theta + $$
$$+ \rho_{33,222222}\sin^2 3\theta)JH^6 , \tag{15}$$

$$E\overrightarrow{[IH]} = \rho_{23,11}\cos 3\theta \, IH^2 + (\rho_{23,1122} + 2\rho_{13,1222})\cos 3\theta \, IH^4 + $$
$$+ R_{23,1}IH + R_{23,111}IH^3 + (R_{23,11111}\cos^2 3\theta + $$
$$R_{31,22222}\sin^2 3\theta)IH^5 . \tag{16}$$

It is evident from Eqs. (15) and (16) that the positions of the extremal values of $R(\Theta)$ and $\Delta\rho/\rho(\Theta)$ depend on the numerical values of the components $R_{23,11111}$, $R_{31,22222}$, $\rho_{33,11111}$, and $\rho_{33,22222}$: a) when $|R_{23,11111}| > |R_{31,22222}|$, and $\rho_{33,11111} > \rho_{33,22222}$, $R(\Theta)$ and $\Delta\rho/\rho(\Theta)$ should have maxima when H is parallel to one of the binary axes; b) if the inequalities are reversed, minima are observed in the same position; c) finally, when the components are equal in pairs, the galvanomagnetic effects in type A samples are independent of the orientation of the binary axes. Obviously, the second case applies in bismuth.

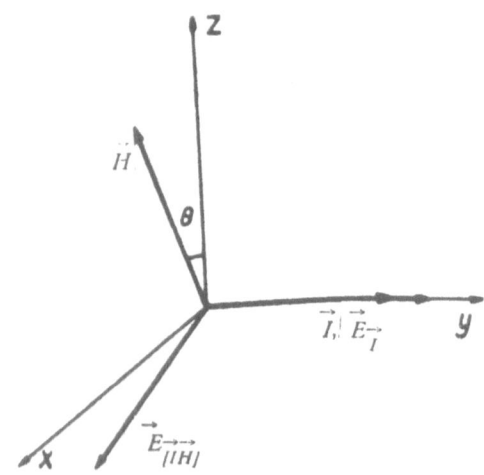

Fig. 12. Type C sample. The orientation of the current $\vec{I}$, magnetic field $\vec{H}$, and electric components $E_{\vec{I}}$ and $E_{[\vec{I}\vec{H}]}$ with respect to the crystallographic axes.

The components $E_{\vec{I}}$ and $E_{[\vec{I}\vec{H}]}$ are, respectively:

Analysis of our data thus shows that if we include a sufficient number of the terms in the expansion of the function $P_{ij}(\vec{H})$, we can obtain analytic expressions for $R(\Theta)$ and $\Delta\rho/\rho\,(\Theta)$, which reproduce the experimental curves. The dependence on H coincides with the observed dependence only in weak magnetic fields.

Type B Samples

It is evident from Fig. 11 that, in the process of rotation of type B samples, the orientation of $\vec{H}$ lies in the yoz-plane. We shall denote the angle between the trigonal axis and $\vec{H}$ by Θ, and regard the $[\vec{I}\vec{z}]$ direction as positive to obtain agreement with the experimental diagrams. Then

$$
\begin{aligned}
I_1 &= I, & H_1 &= 0, \\
I_2 &= 0, & H_2 &= -H\sin\Theta, \\
I_3 &= 0, & H_3 &= H\cos\Theta.
\end{aligned}
\tag{17}
$$

$$
E_{\vec{I}} = E_1 ,
\tag{18}
$$

$$
E_{[\vec{I}\vec{H}]} = -E_2\cos\Theta - E_3\sin\Theta.
\tag{19}
$$

We obtain, after substitution,

$$
\begin{aligned}
E_{\vec{I}} &= \rho_{11}I + (\rho_{11,22}\sin^2\Theta - 2\rho_{11,23}\sin\Theta\cdot\cos\Theta + \\
&\quad + \rho_{11,33}\cos^2\Theta)IH^2 + (\rho_{11,2222}\sin^4\Theta - 4\rho_{11,2223}\sin^3\Theta\cos\Theta + \\
&\quad + 6\rho_{11,2233}\sin^2\Theta\cdot\cos^2\Theta - 4\rho_{11,2333}\sin\Theta\cdot\cos^3\Theta + \\
&\quad + \rho_{11,3333}\cos^4\Theta)IH^4, \\
E_{[\vec{I}\vec{H}]} &= (R_{23,1}\sin^2\Theta + R_{12,3}\cos^2\Theta)IH + [R_{23,111}\sin^4\Theta + \\
&\quad + (R_{12,112} + 3R_{23,123})\sin^3\Theta\cos\Theta + 3(R_{12,113} + \\
&\quad + R_{23,133})\sin^2\Theta\cdot\cos^2\Theta + R_{12,333}\cos^4\Theta]IH^3.
\end{aligned}
\tag{20}\tag{21}
$$

It follows from Eq. (20) that irrespective of the magnetic field intensity, the extremal values of $\Delta\rho/\rho(\Theta)$ are displaced from the $\Theta = 0°$ and $90°$ positions only if the components $\rho_{11,23}$, $\rho_{11,2223}$, and $\rho_{11,2333}$ are not equal to zero. This displacement increases with the magnetic field intensity. It is also easy to understand the replacement of a maximum with a plateau in strong fields. This is obviously due to the terms $6\rho_{11,2233}\sin^2\Theta\cdot\cos^2\Theta$. We can also explain easily and fully the Hall coefficient anisotropy, which can be checked by the comparison of Eq. (21) with the experimental curves. Consequently, to describe qualitatively the curves $R(\Theta)$ and $\Delta\rho/\rho(\Theta)$ of B-type samples, we need only terms containing H^3 and H^4.

Type C Samples

The directions of the current and magnetic field with respect to crystallographic axes in C-type samples are shown in Fig. 12. As in the preceding case, we shall regard the $[\vec{I}\,\vec{z}]$ direction as positive. Then:

$$
\begin{aligned}
I_1 &= 0, & H_1 &= H\sin\Theta, \\
I_2 &= I, & H_2 &= 0, \\
I_2 &= 0, & H_3 &= H\cos\Theta.
\end{aligned}
\tag{22}
$$

Finally,

$$
\vec{E_I} = E_2 = \rho_{11}I + (\rho_{11,22}\sin^2\Theta + \rho_{11,33}\cos^2\Theta)\,IH^2 +
$$

$$
+ (\rho_{22,1111}\sin^4\Theta + 6\rho_{11,2233}\sin^2\Theta\cdot\cos^2\Theta + \rho_{11,3333}\cos^4\Theta)\,IH^4,
\tag{23}
$$

$$
\vec{E}_{[I H]} = E_1\cos\Theta - E_3\sin\Theta = (2\rho_{11,23}\sin\Theta\cdot\cos^2\Theta -
$$
$$
- \rho_{23,11}\sin^3\Theta)\,IH^2 + [3(\rho_{11,1123} + \rho_{11,2223} - 2\rho_{23,1133})\sin^3\Theta\cdot\cos^2\Theta +
$$
$$
+ 4\rho_{11,2333}\sin\Theta\cos^4\Theta - (\rho_{23,1122} + 2\rho_{13,1222})\sin^5\Theta]\,IH^4 +
$$
$$
+ (R_{23,1}\sin^2\Theta + R_{12,3}\cos^2\Theta)\,IH + [R_{23,111}\sin^4\Theta + 3(R_{12,113} +
$$
$$
+ R_{23,133})\sin^2\Theta\cdot\cos^2\Theta + R_{12,333}\cos^4\Theta]\,IH^3.
\tag{24}
$$

It follows from Eqs. (23) and (24) that in weak magnetic fields the extremal values of $R(\Theta)$ and $\Delta\rho/\rho(\Theta)$ should occur at $\Theta = 0°$ and $90°$. In strong fields a second minimum appears due to the term $6\rho_{11,2233}\sin^2\Theta\cdot\cos^2\Theta$, irrespective of the position of a minimum of $\Delta\rho/\rho(\Theta)$. This minimum is clearly visible in pure bismuth. It should also be observed in the $R(\Theta)$ dependence in strong fields; a trace of it appears at $H = 18{,}000$ Oe in pure bismuth.

Thus, for type C samples also the experimentally observed anisotropy of the galvanomagnetic effects is well reproduced by the theoretical expressions.

It follows from the expressions obtained that in samples of types A and C the field $\vec{E}_{[IH]}$ contains both odd and even powers of H. In the method used by us to measure the Hall field, the even powers of H disappear due to the averaging of the experimental results so that the $R(\Theta)$ curves given in the figures have only the odd powers of H.

If we restrict ourselves to terms not higher than H^3, we can, in principle, determine all the components of the tensors ρ_{ij}, $R_{ij,k}$, and $R_{ij,klm}$, and six components of the tensors $\rho_{ij,kl}$, using the three types of sample. We calculated only the components of $\rho_{ij}, \rho_{ij,k}$ and some of the components of $\rho_{ij,kl}$, the numerical values of which are listed in Table 2. Unfortunately, the magnetic fields used were not weak and, therefore, it was difficult to estimate the components of the other tensors.

It is evident from Table 2 that the degree of the dependence of the various components of the same tensor on the tellurium content is not the same, in agreement with the electron theory [5].

Analysis of the results obtained shows that in weak magnetic fields tellurium impurities up to 0.5 at. % do not alter basically the form of the $R(\Theta)$ and $\Delta\rho/\rho(\Theta)$ diagrams. In alloys containing more than 0.1% Te, the ratios $R_{23,1} : R_{12,3}$ and $\rho_{33,11} : \rho_{11,22} : \rho_{11,33}$ remain constant (~ 7, 0.64, and 4.3, respectively). This means that on addition of tellurium, in spite of the sharp increase in the degree of the electron-gas degeneracy, the effective electron mass tensor does not change within the experimental error, i.e., the nature of the constant energy surfaces in the k space remains unaffected [5].

Consequently, the disappearance of the angular dependence of $R(\Theta)$ and $\Delta\rho/\rho(\Theta)$ on the orientation of the crystallographic axes in strong magnetic fields observed on addition of tellurium is mainly due to an increase in the electron density and a reduction in their mobility [6].

CONCLUSIONS

1. It is shown that when a sufficient number of terms is included in the expansion of the functions $P_{ij}(\vec{H})$ we can obtain equations which reproduce well the experimental diagrams $R(\Theta)$ and $\Delta\rho/\rho(\Theta)$ for crystals of all three types over the whole range of magnetic fields used in the present work.

2. The nature of the constant energy surfaces of the conduction band of bismuth is not greatly affected by the addition of up to 0.5% Te, although the Fermi level position is displaced by 15 kT.

LITERATURE CITED

1. D. V. Gitsu and G. A. Ivanov, Fiz. Tverd. Tela 2: 1457 (1960).
2. D. V. Gitsu and G. A. Ivanov, Fiz. Tverd. Tela 2: 1464 (1960).
3. H. I. Juretsche, Acta Cryst. 8: 716 (1955).
4. T. Ocada, J. Phys. Soc. Japan 12: 12 (1957).
5. D. V. Gitsu and G. A. Ivanov, Izv. Akad. Nauk Moldav.SSR, No. 5: 83 (1962).
6. G. A. Ivanov, Fiz. Tverd. Tela 1: 1600 (1959).

PHOTOELECTRIC AND OPTICAL PROPERTIES OF THIN GaTe FILMS

V. I. Gramatskii and V. P. Mushinskii

It is known from the phase diagram [1] that in the Ga−Te system two compounds are distinguished by their melting point: Ga_2Te_3 (73.3 wt. % Te) and GaTe (64.7 wt. % Te). The physical properties of bulk samples and thin films of Ga_2Te_3 have been described in several papers [2-5], whereas the properties of GaTe have been studied relatively litte [2, 6]. It has been found that GaTe in bulk is a semiconductor.

The present paper describes the optical and photoelectric properties of thin GaTe films. The results of a study of bulk single-crystal and polycrystalline samples will be published separately.

Thin films for the measurement of the absorption spectrum and of the steady-state photoconductivity were prepared by a method described earlier [4]. The film thickness was measured interferometrically and ranged from 0.3 to 2 μ.

RESULTS OF MEASUREMENTS AND DISCUSSION

The steady-state photoconductivity, absorption, and reflection spectra were investigated using well-annealed GaTe films. The measurements were carried out using a UM-2 type monochromator. The light intensity was determined with a photocell (type FÉSS-U3) and a mirror galvanometer. The slit width was 0.1 mm.

Figure 1 shows the absorption spectra of GaTe (curve 1) and Ga_2Te_3 (curve 2) films. The absorption coefficient was calculated from

$$K = \frac{ln \frac{I_0 - I_R}{I_d}}{d}, \tag{A}$$

where I_0 is the intensity of incident light, I_R the intensity of reflected light, I_d the intensity of light transmitted by the film, and d the film thickness.

Figure 1 indicates that both GaTe and Ga_2Te_3 absorb strongly in the visible range but beginning from $\lambda \approx 800$ mμ the absorption coefficient is small. The sharp drop of curves 1 and 2 and the relatively high absorption coefficient in the region $\lambda < 900$ mμ indicate that we are dealing with the fundamental absorption. The red edge of the fundamental absorption λ_r is 890 mμ for GaTe and 930 mμ for Ga_2Te_3. The value of ΔE_{opt} for GaTe, determined from λ_r, differs a little from sample to sample, the average being $\Delta E_{opt} = 1.42$ eV. For Ga_2Te_3, $\Delta E_{opt} = 1.32$ eV.

It is known that the forbidden band width ΔE_{opt} determined from the red edge of the fundamental absorption represents the value at the test temperature. A study of the absorption spectra therefore gives the temperature dependence of ΔE which is found by recording the absorption curves at various temperatures. One of the present authors has determined such a dependence for Ga_2Te_3 in an earlier paper [4].

Figure 2 gives the optical density spectra of GaTe films at three different temperatures: -183°C, +20°C, and +70°C. The curves were recorded using the method described in [4]. It is evident from Fig. 2 that the curves shift parallel to one another as the temperature is varied. The variation of ΔE_{opt} and then the temperature

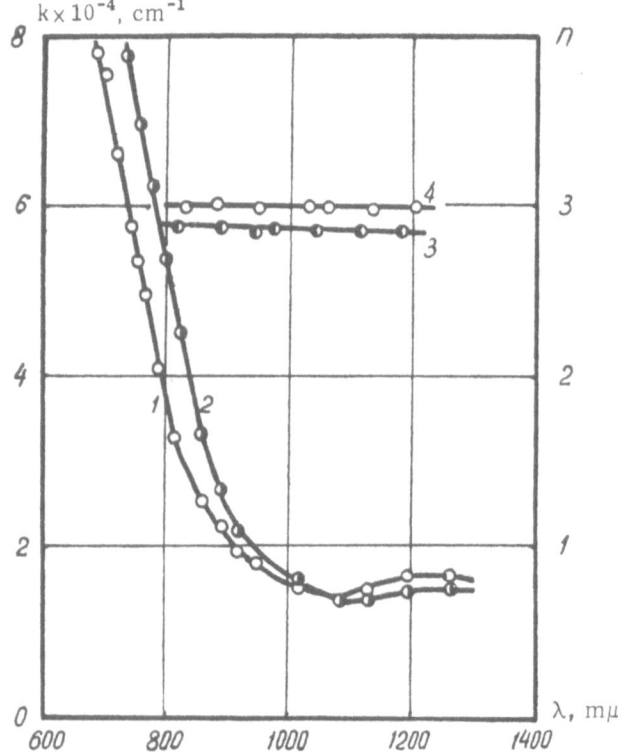

Fig. 1. The absorption spectra and the optical dispersion of GaTe and Ga_2Te_3 films: 1) $k = f(\lambda)$ for GaTe; 2) $k = f(\lambda)$ for Ga_2Te_3; 3) $n = f(\lambda)$ for GaTe; 4) $n = f(\lambda)$ for Ga_2Te_3.

coefficient $\partial E_{opt}/\partial T$ were determined from the variation of λ_r. The temperature dependence of the forbidden band width ΔE_{opt} is shown in Fig. 3. In the temperature range from -183°C to +70°C, the forbidden band width varies linearly with temperature at the rate of $\alpha = \partial E_{opt}/\partial T = -4.8 \times 10^{-4}$ eV/deg. Extrapolating the $\Delta E_{opt}(T)$ curve to 0°K, we find that $\Delta E_0 = 1.52$ eV. Thus, the temperature dependence of the forbidden band width of GaTe, ΔE_{opt}, can be satisfactorily described by

$$\Delta E_{opt} = \Delta E_0 - 4.8 \cdot 10^{-4} T \text{ eV/deg.}$$

Figure 4 gives the reflection spectra of two GaTe films of different thicknesses, which should be compared with the reflection spectra of the two Ga_2Te_3 films in Fig. 5. Figure 4 shows, in the weak absorption region, a series of maxima and minima due to the interference of light in thin GaTe films. For near-normal incidence of light, the reflected light exhibits minina if $2dn = m\lambda$, and maxima if $2dn = [(2m+1)/2]\lambda$, provided $n > n_0$, where n and n_0 are, respectively, the refractive indices of the film and the glass substrate on which it is deposited, m is the order of the interference band, found from two neighboring extrema,

$$m = \frac{1}{2} \cdot \frac{\lambda_2}{\lambda_1 - \lambda_2},$$

where λ_1 and λ_2 are the wavelengths of a minimum and a neighboring maximum in the reflected light. Knowing the order of interference m, we can determine the refractive index of the film

$$n = \frac{1}{4a} \cdot \frac{\lambda_1 \cdot \lambda_2}{\lambda_1 - \lambda_2}.$$

Fig. 2. Temperature dependence of the absorption by GaTe films: 1) t = -183°C; 2) t = 20°C; 3) t = 70°C.

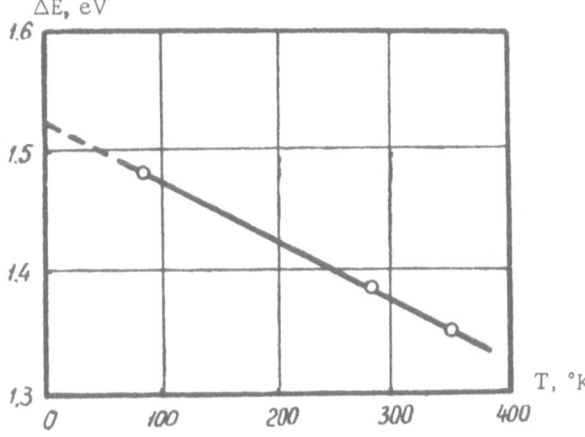

Fig. 3. Temperature dependence of ΔE_{opt} of GaTe.

The optical dispersion for GaTe and Ga_2Te_3 films, determined from interference maxima and minima, is shown in Fig. 1 by curves 3 and 4. It is evident from Fig. 1 that in the weak absorption region the refractive indices decrease with increase of the wavelength. The refractive index of GaTe is somewhat lower than the refractive index of Ga_2Te_3. This difference in the refractive indices is in good agreement with Moss's rule [7] that $\Delta E \cdot n^4 = const$.

GaTe films, like single crystals, exhibit photoconductivity, their sensitivity being relatively higher than that of Ga_2Te_3 films. Photoconducting GaTe films were obtained using the techniques described above. Aquadag served as the electrode material. The steady-state photoconductivity was investigated using white and monochromatic light.

Figure 6 presents the photoconductivity spectra of GaTe (curve 1) and Ga_2Te_3 (curve 2) films. These spectra are approximately similar but their maxima are shifted somewhat with respect to one another. For GaTe $\lambda_{max} = 730$ mμ, which corresponds to a photon energy of 1.7 eV; for Ga_2Te_3 λ_{max} 770 mμ. The positions of the maxima for a large number of samples lie within 100-150 A, which is greater than the experimental error for λ_{max}.

The forbidden band width may be found from the photoconductivity spectra. The best-known method is that of Moss, according to whom the forbidden band width ΔE_{ph} is found from $\lambda_{1/2}$, which is the wavelength at the long-wave edge of the $\Delta I = f(\lambda)$ curve at which the photocurrent is half its maximum value. For GaTe at room temperature, $\lambda_{1/2} = 850$ mμ, which corresponds to $\Delta E_{ph} = 1.45$ eV. This value of ΔE_{ph} is in good agreement with the results obtained by optical methods and from the temperature dependence of the electrical conductivity. The temperature dependence of the forbidden band width may also be found from the photoconductivity spectra. Figure 7 shows the steady-state photoconductivity spectra of GaTe at three different temperatures: -183°C, +20°C, and +70°C. The measurements were carried out in vacuum in a special Dewar vessel. It follows from Fig. 7 that the curves shift monotonically with temperature. The temperature coefficient of the forbidden band width, $\alpha_1 = \partial E_{ph}/\partial T$, found from the shift of $\lambda_{1/2}$ is 4.5×10^{-4} eV/deg. This value of α_1 is in good agreement with the results obtained from the optical measurements. Thus, we can definitely say that the forbidden band width of GaTe decreases linearly as the temperature increases. The nature of this dependence can be described by the following general formula $\Delta E = \Delta E_0 - \alpha T$, where $\alpha = (4.5-4.8) \times 10^{-4}$ eV/deg. The increase of the forbidden band width of GaTe on cooling may be due to some change in the interaction of electrons with the lattice vibrations.

The temperature dependence of the photocurrent of GaTe has not yet been studied at all. We made an attempt to determine this dependence. A knowledge of the temperature dependence of the photocurrent is also

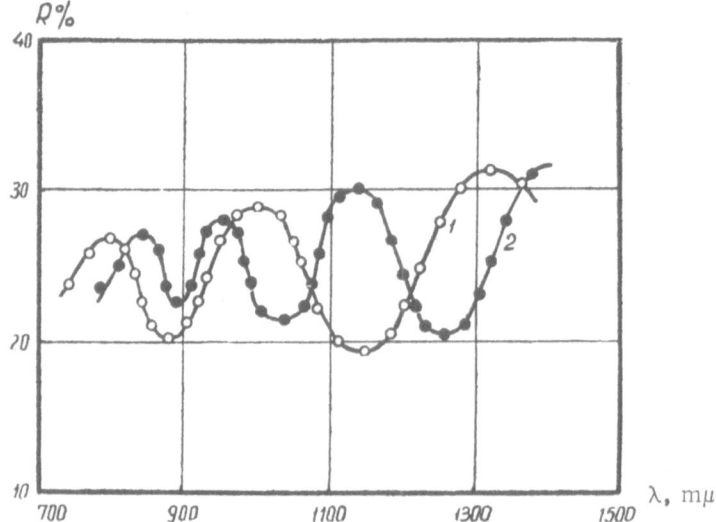

Fig. 4. The reflection spectra of GaTe films of various
thicknesses: 1) $d = 0.6\ \mu$; 2) $d = 1.2\ \mu$.

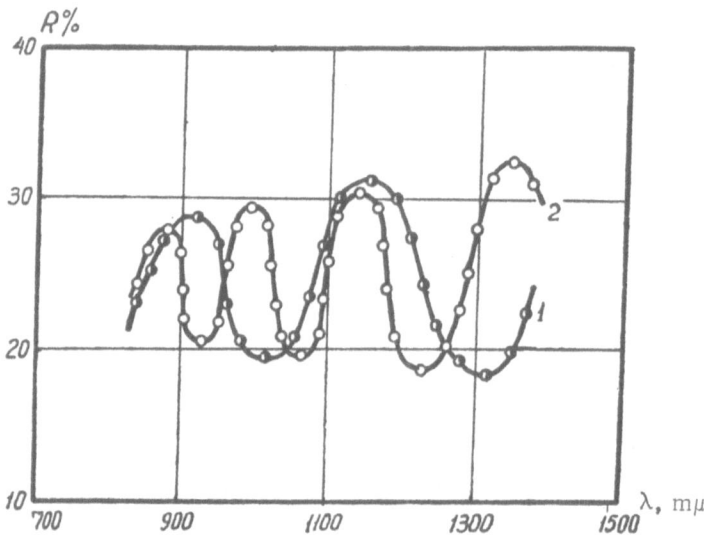

Fig. 5. The reflection spectra of Ga_2Te_3 films of various
thicknesses: 1) $d = 0.75\ \mu$; 2) $d = 1.3\ \mu$.

valuable because it allows us to obtain information on the mechanism of processes occurring in a photoconducting material during illumination. Figure 8 shows the dependence of the dark and photocurrent on temperature. The dark current rises rapidly, beginning from 150°K. The temperature dependence of the photocurrent is shown by curve 2 with a maximum at 190-200°K. At low temperatures, the carrier lifetime is the factor which exerts the greatest influence on the temperature dependence of the steady-state photoconductivity. Ryvkin [8] has shown that in crystals with one type of trap the steady-state photocarrier lifetime is

$$\tau = A T^{3/2} \cdot e^{\frac{\Delta E_M}{kT}},$$

where A is a constant, ΔE_M is the energy separation between the trap levels and the edge of that band whose carriers are in dynamic equilibrium with the trapped electrons. If the photocurrent rise is solely due to the temperature dependence of the cross section for capture by the traps, the experimental curve $\ln(\Delta I / T^{3/2}) = f(1/T)$

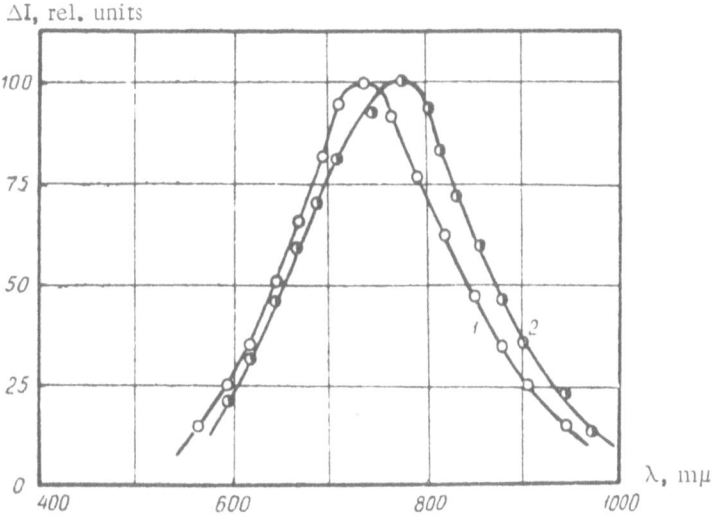

Fig. 6. Photoconductivity spectra of GaTe (curve 1) and Ga_2Te_3(curve 2).

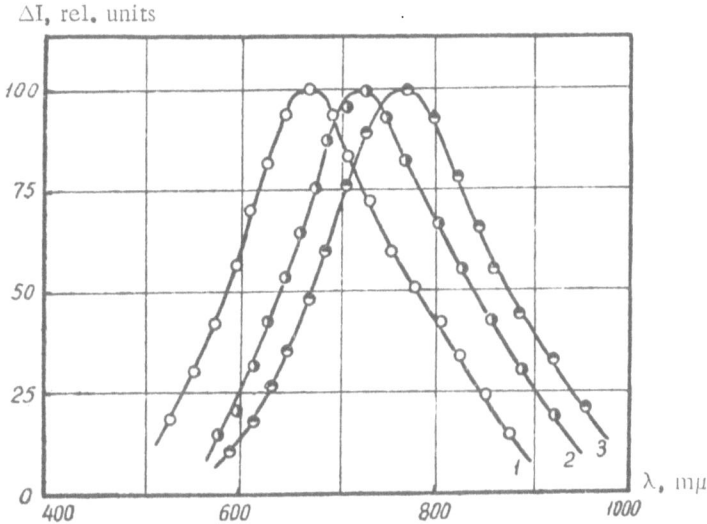

Fig. 7. Temperature dependence of the photoconductivity spectra of GaTe: 1) t= -183°C; 2) t= 20°C; 3) t= 70°C.

should be a straight line. We checked the linearity of this curve in the rising part of $\Delta I = f(T)$. It was found that the experimental points fitted a straight line quite well. From the slope of the line $\ln(\Delta I/T^{3/2}) = f(T)$, the value of ΔE_M was determined: it was found to be of the order of 0.01 eV. Comparison of the value of ΔE_M with the energy of thermal motion, kT, showed that in the temperature range 90-200°K the photocurrent rise may be accounted for by thermal liberation of carriers, localized at the trap levels.

The current−voltage and the illumination−current (lux−ampere) characteristics of photoconductive materials are of great practical and theoretical interest.

Figure 9 gives the current−voltage characteristics for the photocurrent (curve 1) and the dark current (curve 2) in GaTe. The curves show that the photocurrent rises linearly with the voltage. The dark current rises linearly up to V = 300 V, which corresponds to a field intensity $E = 2 \times 10^3$ V/cm. In stronger fields, the dark current rises superlinearly with the voltage. This apparently is due to the heating of the film by the current passing through it.

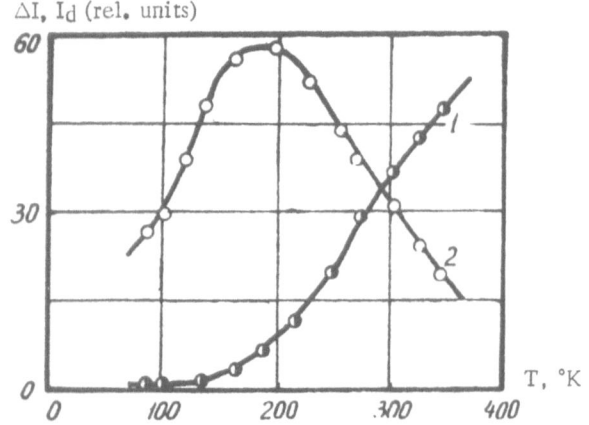

Fig. 8. Temperature dependence of the dark current (curve 1) and photocurrent (curve 2) in GaTe illuminated with white light of E = 5000 lux intensity.

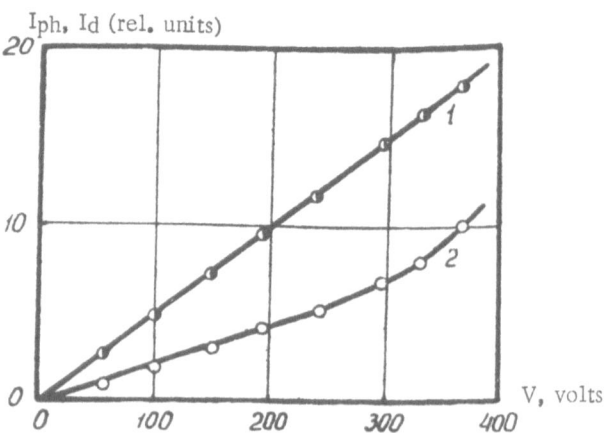

Fig. 9. Current—voltage characteristics of GaTe: 1) photocurrent at E = 5000 lux; 2) dark current.

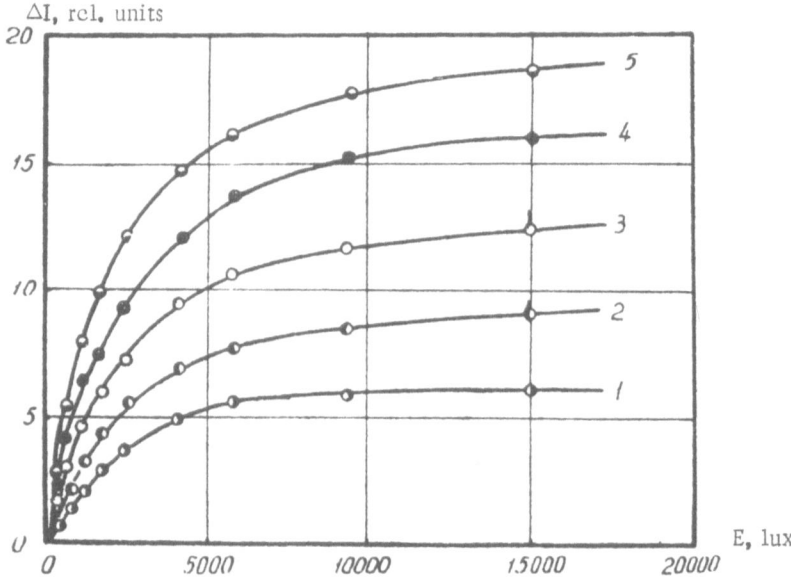

Fig. 10. Dependence of the photocurrent on the illumination of a GaTe sample subjected to the following voltages V: 1) 50 V; 2) 100 V; 3) 160 V; 4) 200 V; 5) 250 V.

The dependence of the photocurrent in GaTe on the illumination is given in Fig. 10. The various curves represent different voltages applied to a sample. The lux—ampere characteristics of GaTe are well approximated by the law $\Delta I = \sqrt{L}$, where L is the incident light intensity, which suggests the bimolecular recombination mechanism.

CONCLUSIONS

1. The absorption, reflection, and steady-state photoconductivity spectra of thin GaTe films were investigated in the visible part of the spectrum. The optical width of the forbidden band ΔE_{opt} of GaTe was found to be 1.42 eV at room temperature.

2. The temperature coefficient of the forbidden band width was $\alpha = -4.8 \times 10^{-4}$ eV/deg (ΔE_{opt} decreased linearly with increase of temperature).

3. The refractive index of GaTe films outside the fundamental absorption band was of the order of 2.7 and increased on reduction of the wavelength.

4. The photocurrent spectrum of GaTe exhibited one maximum at $\lambda_{max} = 730$ mμ in the absorption band. The forbidden band width, ΔE_{ph}, found from $\lambda_{1/2}$ for the photocurrent, was 1.45 eV.

5. On increase of temperature, the spectra shifted toward longer wavelengths. The temperature coefficient of the shift of ΔE_{ph} was -4.5×10^{-4} eV/deg.

6. The depth of trap levels in GaTe was ≈ 0.01 eV.

7. The steady-state photoconductivity maximum occurred at t = -70 to -80°C.

LITERATURE CITED

1. W. Klemm and H. U. Vogel, Z. Anorg. u. Allgem. Chem. 219: 45 (1934).
2. N. A. Goryunova, V. S. Grigor'eva, B. M. Konovalenko, and S. M. Ryvkin, Zh. Tekhn. Fiz. 25(10) (1955).
3. G. Harbeke and G. Lautz, Z. Naturforsh. 13a: 775 (1958).
4. V. I. Gramatskii, Uch. Zap. Kishnievsk. Gos. Univ. 49: 119 (1961).
5. V. P. Mushinskii, V. I. Gramatskii, and G. N. Manushevich, Izv. Vuzov MVO, Fizika, No. 3: 172 (1963).
6. P. Fielding, G. Fischer, and E. Mosser, J. Phys. Chem. Solids 8: 434 (1959).
7. T. Moss, Optical Properties of Semiconductors [Russian translation], Moscow (1961).
8. Semiconductors in Science and Technology, Vol. 2 (1958), p. 463.

THE CRYSTALLIZATION OF SEMICONDUCTORS
FROM A MOLTEN SOLUTION

G. A. Kalyuzhnaya, D. N. Tret'yakov, A. S. Borshchevskii,
and A. A. Vaipolin

A review of the literature on the practical use of semiconducting phases with tetrahedral atomic coordination reveals the following features.

First, the range of applications of these materials increases with their position in the periodic system. This is because of the need for instruments working at high temperatures, for hard-radiation receivers, etc.

Secondly, in those applications where until recently elemental semiconductors and binary compounds were used almost exclusively, more complex semiconducting materials, including solid solutions, are now employed. This is because various multicomponent materials have better (compared with elemental and binary semiconductors) combinations of properties.

At present, the range of practical applications of refractory materials is greatest in the fourth group. Many instruments using silicon carbide have already been tested, and work has appeared describing studies of the semiconducting properties of diamond. The range of applications of other semiconducting materials with tetrahedral coordination is narrower. For example, in the case of $A^{III}B^{V}$-type compounds the limits of the range of applications pass close to GaP, but exclude the promising compounds of arsenic and phosphorus with aluminum and boron, as well as nitrogen compounds.

To prepare and study these substances, it is necessary to select the most convenient method of synthesis.

Recently, one of the "direct" methods of making semiconductors was employed for this purpose [2, 5, etc.]. This method consists of the preparation of refractory materials by their crystallization from molten solutions using an easily fusible solvent. In some cases, this method has certain advantages over indirect methods. Since this direct method is more universal than the indirect methods, it has become possible to prepare highly pure materials in the form of single crystals. The high purity is due to relatively low temperatures during the preparation and to large amounts of extremely pure easily fusible solvent. This method also makes it possible to alloy the material being prepared [1].

The majority of papers on the use of this method have dealt with the preparation of elemental semiconductors in the fourth group of the periodic system [2-4] and of some $A^{III}B^{V}$-type compounds, namely, InSb, InAs, InP, GaSb, GaAs, GaP, AlSb [5, 21], BP [6].

There is great interest in the preparation of some other compounds by the same method, in particular solid solutions [22, 23].

In the present work, the method of crystallization from a molten solution was used to prepare some binary compounds and solid solutions based on them.

TABLE 1

Solid solution type	Substance	Investigated solid solution range*	Previous published work
Isovalent anion substitution	GaSb—GaAs	0—10 mol. % GaAs	Suggestion of the formation of solid solutions at all concentrations [14]
"	GaSb—GaP	< 5 mol. % GaP	Not investigated earlier
"	GaSb—GaP	All concentrations	Solid solutions were obtained [15]
Isovalent cation	GaSb—AlSb	Wide range of concentrations	Solid solutions formed at all concentrations [16]
"	GaAs—AlAs	0-30 mol. % AlAs	Solid solutions formed in a wide range of concentrations [17]
"	GaP—AlP	0-50 mol. % AlP	Solid solutions not investigated earlier
Heterovalent mixed	GaP—ZnSe	About 50 mol. %	Not investigated earlier
"	GaP—ZnS	Wide range of concentrations	Not investigated earlier
"	GaP—CdS	Solid solutions not found along this tie-line	Not investigated earlier

*All calculations were based on the assumption that Vegard's law was obeyed.

EXPERIMENTAL METHOD

Our solvent was gallium since it allowed us to investigate the crystallization of a large number of interesting substances without affecting their properties. Moreover, the low melting point of gallium simplified considerably the technique of separating grown crystals.

We used dilute solutions, in accordance with the method described earlier [7].

We did not attempt to prepare equilibrium solutions and, therefore, the rates of cooling during synthesis were relatively high. About 150 melting operations were carried out. The crystals obtained were investigated mainly by x-ray diffraction. In some cases, we measured the microhardness and studied the microstructure.

RESULTS AND DISCUSSION

The necessary stage in the preparation of solid solutions was the synthesis of several binary components used to form these solutions, namely, GaSb, GaAs, GaP, AlSb, AlAs, AlP, ZnSe, ZnS, CdS. In the case of GaAs and GaP, we were able to prepare single-crystal plates convenient for electrical studies.

Single crystals of GaP were of special interest because this material has not been studied much although it seems very promising. The GaP single crystals were in the form of transparent orange tinted plates measuring up to $9 \times 5 \times 0.6$ mm. The lattice parameter, measured with an RKU-114 type camera (using Cu K_α radiation), was found to be 5.439 ± 0.001 kXU (ZnS structure), in good agreement with the published data [8]. The microhardness was measured with a PMT-3 type instrument (using a square-based pyramid): it was 960 ± 20 kg/mm^2. The density measured by the method described in [9] was 4.10 g/cm^3, compared with 4.15 g/cm^3 calculated from x-ray diffraction data.

The GaP crystals were of n- or p-type containing 2×10^{17} cm^{-3} electrons having a mobility of 130 cm$^2 \cdot$ V$^{-1} \cdot$ sec^{-1} at room temperature [10]. The mobility found for our samples was greater than the electron mobility reported in the published work [11]. The absorption spectrum of single-crystal gallium phosphide has also been investigated [12].

The next stage was the preparation of binary compounds of Al. There are published data on the crystallization of only one of these compounds — AlSb [5], made using Al as a solvent. We used gallium as the solvent because it did not form the compounds with Al. That might have led to the solution of a considerable amount of Ga in the aluminum compounds, if the heterogeneous equilibrium GaX$_{sol}$ + Al$_{liq}$ $\leftrightarrows$ AlX$_{sol}$ + Ga$_{liq}$ had not shifted completely to the right during crystallization. Then, instead of the compund AlX, the solid solutions AlX — GaX might have been formed, with the likely appearance of aluminum in the solvent. Precisely this occurred during the crystallization of AlSb from Ga. Accurate measurements of the lattice parameter showed that the solid solution AlSb — GaSb, and not the compound AlSb, crystallized out from the melt.

The crystallization of the compounds AlAs and AlP from gallium differed from the crystallization of AlSb. Pure compounds and not solid solutions were obtained. This was likely to be due to the greater difference between the heats of formation of AlP and GaP, AlAs and GaAs, than the difference between the heats of formation of AlSb and GaSb. There has been very little work on the refractory compounds AlAs and AlP. In our tests, AlAs was obtained in the form of transparent dark red crystals, and AlP in the form of weakly colored yellow crystals, which indicated low concentration of impurities, in good agreement with the forbidden band width of these compounds (> 2 eV). This was expected for the gallium solvent because this element is not an electrically active impurity in compounds of the A^{III}B^{V} type. Precise measurements of the lattice parameter showed that gallium was not dissolved in noticeable amounts in these compounds. The lattice parameter of AlAs measured by us (a = 5.648 kXU) was in agreement with the published value [8]. The published value of the lattice parameter of AlP [13] was not very accurate and consequently we could not compare it with our value to provide a convincing proof of the absence of the solubility of gallium in AlP. Therefore, we carried out more precise measurements of the lattice parameter of AlP using a sample prepared by the synthesis of aluminum and phosphorus in stoichiometric proportions, using a method described earlier [7]. It was found that the lattice parameter of AlP prepared in this way was 5.452 ± 0.001 kXU, while the parameter for AlP prepared by crystallization from a solution in gallium was 5.451 ± 0.001 kXU.

Apart from the compounds A^{III}B^{V}, we also investigated A^{II}B^{VI}-type compounds. Crystals of ZnS and ZnSe prepared by crystallization from gallium did not differ in color or in x-ray structure from samples of the same compounds prepared by other methods. The problem of the influence of small amounts of gallium on the electrical properties of ZnS and ZnSe requires a special study. We were unable to prepare CdS by crystallization from gallium because the equilibrium, such as that mentioned above, shifted completely in the direction of compounds of sulfur with gallium. The results of our study of the formation of solid solutions are collected in Table 1.

This table shows that almost all systems (except AlSb — GaSb) contain at least one refractory component. The following points are worth mentioning in connection with the crystallization from the melt. First, the data on solid solutions were obtained for the stoichiometric tie-lines listed in Table 1. Our investigations of the crystallization of binary compounds from molten gallium confirmed that in practice there were no departures from stoichiometry. This was in agreement with the published data on the very narrow range of homogeneity in binary compounds with the ZnS structure [18].

Secondly, the limits of the solid solution formation quoted here are preliminary. These limits indicate that the method of crystallization from molten gallium does not restrict the solubility limits. These limits depend on the concentration, temperature, and kinetic conditions, and even on the method of separation. We mention here the interdependent concentration and temperature conditions separately in order to stress one of the features of the crystallization from molten solutions (using an easily fusible solvent) as compared with the crystallization of solid solutions from the melt with the components in stoichiometric proportions. If, according to the phase rule, a system is univariant in the stoichiometric case, then in the method of crystallization from molten solutions using an easily fusible component the number of degrees of freedom is greater and the solid phase of a given composition may crystallize at various temperatures. Consequently, it may happen that in some variants we are below the theoretically predicted temperature of the onset of precipitation in a solid solution [19].* This particular circumstance imposes certain temperature and concentration restrictions on the solid solutions being prepared. This is of scientific interest because it allows us to use the method of crystallization from molten solution in an experimental verification of theoretical hypotheses.

We shall consider in some detail the results on certain particular systems. Among systems with anion isovalent substitution, the results for the $Ga-Sb-As$ system are of special interest, in spite of the relatively restricted solubility (10.5 mol. % GaAs in solid solution). Apart from the general prediction of solubility over the whole range of concentrations and the promise that the results will be published later [14], there are no published data on this point.

In the similar system $In-Sb-As$, homogeneous samples are obtained by the annealing of powders. In the GaSb$-$GaAs system, homogeneous solid solutions can be obtained only in the form of powders if the usual synthesis methods are employed. Our data on the crystallization of stoichiometric melts of composition along the pseudobinary tie-line GaAs$-$GaSb confirmed these reports. The solid solution range was less than 1% GaAs in GaSb. With crystallization from gallium, we at once obtained solid solutions with much higher solubility, which we raised to 10.5% GaAs by reducing the cooling rate. Further investigations should lead to the preparation of samples homogeneous over a wider range of concentrations and suitable for physico-chemical and electrical measurements.

It is known that the binary compounds AlAs and AlP exhibit corrosion instability. Therefore, it is very significant that a slight solubility in HCl was detected for solutions with 50 mol. % in the GaP$-$AlP system and 30 mol. % AlAs solutions in the GaAs$-$AlAs system. One therefore hopes that corrosion-stable alloys may be obtained which are close in their composition and properties to AlAs and AlP.

The solid solutions along the CdS$-$GaP tie line were not obtained. As expected, solid solutions were formed along the GaP$-Ga_2S_3$ tie line, in full agreement with the results obtained on crystallization of CdS from Ga.

The experimental results reported here and some considerations of the crystallization from molten solutions suggest that this method is a very promising one.

LITERATURE CITED

1. R. Solomon, J. Electrochem. Soc. 108: 716 (1961).
2. P. H. Keck and J. D. Bröder, Phys. Rev. 90: 521 (1953).
3. O. N. Tufte and A. W. Ewald, J. Appl. Phys. 29: 1007 (1958).
4. F. A. Trumbore, J. Phys. Chem. Solids (1962).
5. G. Wolff, P. H. Keck, and J. D. Bröder, Bull. Am. Phys. Soc. 29: 16 (1954).
6. Yu. A. Valov and É. Yu. Lubenskaya, Proceedings of the Twentieth Scientific Conference of the Leningrad Building Institute, Physics Section, Leningrad (1962), p. 31.
7. A. S. Borshchevskii and D. N. Tret'yakov, Fiz. Tverd. Tela 1: 1483 (1959).
8. G. Giesecke and H. Pfister, Acta Cryst. 11: 369 (1958).
9. M. M. Vasilevskii, Zavodsk. lab., No. 10: 1170 (1960).
10. D. N. Nasledov and S. V. Slobodchikov, Fiz. Tverd. Tela 4(10) (1962).
11. C. Hilsum and A. C. Rose-Innes, Semiconducting III$-$V Compounds, Oxford (1961).
12. E. F. Gross, G. A. Kalyuzhnaya, and D. S. Nedzvetskii, Fiz. Tverd. Tela 3: 3543 (1961).
13. H. G. Grimmeiss, W. Kischio, and A. Rabenau, J. Phys. Chem. Solids 16: 302 (1960).
14. J. C. Woolley and B. A. Smith, Proc. Phys. Soc. (London) 72: 214 (1958).
15. O. G. Folberth, Z. Naturforsch. 10a: 502 (1955).
16. N. A. Goryunova and I. I. Burdiyan, Dokl. Akad. Nauk SSSR 120(5) (1958).
17. E. P. Stambaugh, I. I. Genco, and R. C. Himes, J. Electrochem. Soc. 107: 91 (1960).
18. N. A. Goryunova, Chemistry of Diamond-Like Semiconductors, LGU, Leningrad (1962).
19. V. N. Romanenko and V. I. Ivanov-Omskii, Dokl. Akad. Nauk SSSR 129: 559 (1959).

---

*The calculations reported in [19] have not yet been checked experimentally because of the low diffusion rates at the precipitation temperatures in these systems.

20. K. Makoto and S. Iizima, J. Phys. Soc. Japan 16: 1783 (1961).

21. N. A. Goryunova, A. S. Borshchevskii, Yu. A. Valov, G. A. Kalyuzhnaya, and D. N. Tret'yakov, Certificate of Registration of Scientific Research Work No. 18827 with priority date June 3, 1960.

22. E. P. Stambaugh, J. E. Miller, and R. C. Himes, Metallurgy of Elemental and Compound Semiconductors, Interscience, New York, (1961), pp. 317-327.

SOME RESULTS OF A STUDY
OF THE MICROHARDNESS ANISOTROPY OF BISMUTH

T. I. Lange

One of the currently favored methods of physico-chemical analysis is the study of microhardness [1-4]. Microhardness is a very convenient but highly complex characteristic depending on the distribution of atoms in the crystal lattice, the nature of the chemical bonds, the interatomic spacing, the surface state, and several other factors. An exact quantitative relationship between these factors and the microhardness has not yet been established and only some interesting qualitative correlations have been found [5-13]. Nevertheless, the study of microhardness is now widely used in the laboratory practice both in investigations of elemental semiconductors [6-10, 14, 15] and complex semiconductor systems [5, 12, 13, 16-18].

Recently, the microhardness method has been used to study the anisotropy of the mechanical properties of substances with covalent binding [14, 19]. Ablova and Regel' [14] investigated the anisotropy of germanium by the indentation method; the same property of tellurium [19] was investigated by the scratch method [20], which has also been applied successfully to ionic crystals [21, 22].

From very general considerations, it follows that the hardness rosettes obtained by the scratch method should reflect the bond anisotropy in the lattice of the investigated substance. It seems that the results obtained in the study of the microhardness of tellurium [19] confirm this conclusion.

The present paper describes briefly some results of a study of the microhardness anisotropy of bismuth, which, like tellurium, exhibits a strong bond anisotropy.

Bismuth crystals belong to the trigonal system (D_{3d}^5 class [23]) and have a layered structure. Atoms of each layer lie in two parallel planes in such a way that any atom in one plane has three nearest neighbors in the other plane of the same layer and three further neighbors in the next layer. The binding between atoms in one layer is mainly covalent, while the neighboring layers are bound by forces partly of the van der Waals type and partly of the metallic type.

Single crystals grown by the Bridgman method [24] from bismuth of V-000 grade were split along perfect cleavage planes (0001) and investigated using the method described earlier [19].

Figure 1 gives the angular dependence of the hardness, calculated from the formula,

$$H = \frac{P}{d^2},$$

where d is the scratch width and P is the load on the pyramid of a PMT-3 type instrument; the angle in Fig. 1 gives the scratch direction. This figure shows that the hardness rosette has threefold symmetry, identical with the symmetry of the (0001) face. However, in contrast to the results obtained for the (0001) face of tellurium (cf. Fig. 9 in [19]), the angular dependence of the hardness of bismuth is much more complex: there is a weak maximum between two strong maxima separated by 120°. The depths of the minima on the two sides of the weak maximum are not equal: the first is 7 kg/mm², and the second is 8 kg/mm².

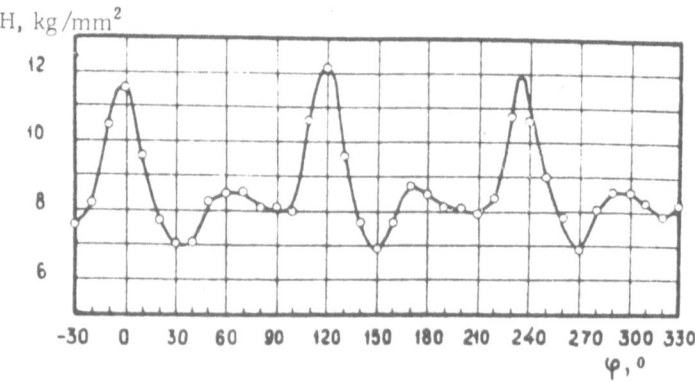

Fig. 1. Dependence of the scratch microhardness $H = P/d^2$ on the scratch direction in the basal plane (0001) of pure bismuth.

The highest values of the hardness, observed when the scratch is perpendicular to the directions of emergence of secondary cleavage planes on the trigonal (basal) plane, may be explained by the breaking of relatively strong bonds in these planes. The origin of the intermediate hardness maximum is not clear.

The form of the dependence of the microhardness on the scratch direction on the (0001) face is well represented by the expression

$$H = A + B\cos3\varphi + C\sin^2 3\varphi,$$

which is similar to the formula describing the Hall effect anisotropy in bismuth crystals with the trigonal axis directed along the sample axis [25]. It is likely that this similarity is not accidental but is of basic importance, indicating the same cause of the anisotropy in both cases.

In conclusion, the author expresses his gratitude to M. S. Ablova, D. V. Gitsu, T. A. Kontorova, V. N. Lange, and A. R. Regel' for their advice and their readiness to discuss the work.

LITERATURE CITED

1. M. M. Khrushchev and E. S. Berkovich, PMT-2 and PMT-3 Instruments for Microhardness Tests, Izd. Akad. Nauk SSSR (1950).
2. M. M. Khrushchev, Microhardness, in: Proceedings of a Conference on Microhardness, November 21-23, 1950, Izd. Akad. Nauk SSSR, Moscow (1951).
3. V. M. Glazov and V. N. Vigdorovich, Microhardness of Metals, Moscow (1962).
4. B. W. Mott, Micro-Indentation Hardness Testing [Russian translation], Moscow (1960).
5. C. C. Wang and B. H. Alexander, in the collection: Silicon [Russian translation] Moscow (1960).
6. V. N. Lange and A. R. Regel', Fiz. Tverd. Tela 1: 559 (1959).
7. J. H. Westbrook, J. Electrochem. Soc. 103: 54 (1956); 104: 369 (1957).
8. D. V. Gitsu, G. A. Ivanov, and V. G. Luzhkovskii, Problems of physics of semiconductors and dielectrics, Uch. Zap. Lenigr. Gos. Ped. Inst. im. A. I. Gertsena 207: 45 (1961).
9. G. C. Kuczynski and R. L. Hochman, Phys. Rev. 108: 946 (1957).
10. M. S. Ablova and A. R. Regel', Fiz. Tverd. Tela 4: 1053 (1962).
11. T. A. Kontorova, in collection: Some Problems of the Strength of Solids, Izd. Akad. Nauk SSSR, Moscow (1959), p. 99.
12. V. P. Zhuze and T. A. Kontorova, Zh. Tekhn. Fiz. 28: 1727 (1958).
13. A. S. Borshchevskii, N. A. Goryunova, and N. K. Takhtareva, Zh. Tekhn. Fiz. 27: 1408 (1957).
14. M. S. Ablova, Fiz. Tverd. Tela 3: 1815 (1961).
15. M. S. Ablova, Fiz. Tverd. Tela 3: 3133 (1961).
16. N. A. Goryunova, S. I. Radautsan, and V. I. Deryabina, Fiz. Tverd. Tela 1: 513 (1959).

17. I. I. Burdiyan and A. S. Borshchevskii, Zh. Tekhn. Fiz. 28: 2684 (1958).

18. N. A. Goryunova, S. I. Radautsan, and G. A. Kiosse, Fiz. Tverd. Tela 1: 1858 (1959).

19. Yu. S. Boyarskaya, V. N. Lange, and T. I. Lange, this volume, p. 39.

20. V. D. Kuznetsov, Surface Energy of Solids, GITTL, Moscow (1954).

21. Yu. S. Boyarskaya, Uch. Zap. Kishinevsk. Univ. 17: 159 (1955).

22. Yu. S. Boyarskaya, Kristallografiya 2: 709 (1957).

23. B. F. Ormont, Structure of Inorganic Substances, GITTL, Moscow (1950).

24. H. E. Buckley, Crystal Growth [Russian translation], IL, Moscow (1954).

25. D. V. Gitsu, and G. A. Ivanov, this volume, p. 60.

MICROCHEMICAL PHASE ANALYSIS OF
SOME SEMICONDUCTING ALLOYS OF THE In—Sb—Te SYSTEM

Yu. S. Lyalikov, L. S. Kopanskaya, I. P. Molodyan, and S. I. Radautsan

In their analyses of semiconducting materials, chemists have been primarily interested in the determination of trace amounts of impurities which affect the semiconducting properties. In the investigations of complex semiconducting alloys, there is also great interest in the determination of the principal components of the alloys. However, so far the techniques for the determination of the chemical composition have been developed for a relatively small number of multicomponent systems. Even less developed is the local phase analysis of complex semiconducting systems. X-ray diffraction, microstructure analysis, thermal analysis, and measurements of the microhardness give information on the structure and properties of a substance as a whole or its various phases, and they are used as indirect methods of determining the composition from the known phase diagrams. However, these methods do not always give the true chemical compositon of a phase with sufficient accuracy.

The necessity for chemical phase analysis appears particularly in the study of the homogenization of alloys by various methods, expecially by zone melting. The chemical phase analysis of complex semiconducting substances prepared by zone melting is used, together with other methods, to find the efficiency of zone melting and to give additional information on the processes occurring during homogenization. All this has led us to attempt the chemical phase analysis of semiconducting alloys based on indium, antimony, and tellurium.

The earlier investigations of the physico-chemical properties of alloys of the ternary system indium—antimony—tellurium [1-3] have shown that solid solutions with the zinc blende structure are formed along the tie-lines $In_3Sb_3-In_2Te_3$ and $InSb-InTe$ in the region 0-15 equimol. % In_2Te_3 and $InTe$. A ternary compound denoted by the formula In_4SbTe_3, with the rock-salt structure, has been discovered along the $InSb-InTe$ tie-line. The data from microstructure, x-ray structure, and thermal analyses have shown the presence of several phases in alloys containing less that 85 equimol. % $InSb$. The differences appearing in the structure, microhardness, melting point, and other properties of these phases can only hint at their actual chemical composition. Therefore, in studies of the phase composition of complex semiconducting systems — in particular, the indium—antimony—tellurium system — the direct microchemical analysis methods are of very great interest.

In an earlier work [4], we showed that chemical analysis methods developed for the indium—antimony—tellurium system [5] are not suitable for the analysis of samples weighing less than 0.1 mg. Therefore, an attempt was made to use physico-chemical analysis methods [6] for this purpose. The present paper reports the first results on the microanalysis of the phase composition of the In—Sb—Te system, obtained by the potentiometric method. Antimony was determined by oxidation by bromate-bromide in an acid medium, while tellurium was determined by reduction with potassium iodide followed by titration of the liberated iodine with sodium thiosulfate [6]. The nature of the titration curves of antimony and tellurium was described earlier [4].

To determine indium we used the formation of a complex with sodium ferrocyanide in a weakly acid medium [7, 8].

To determine indium in an alloy, the latter was dissolved first by the method described in [5]. From the solution, an aliquot part (1 ml) was taken diluted with water to 20 ml and titrated cold with sodium ferrocyanide. Deichman and Tananaev [7] suggested that a sharp jump of the potential should occur at the equivalent point.

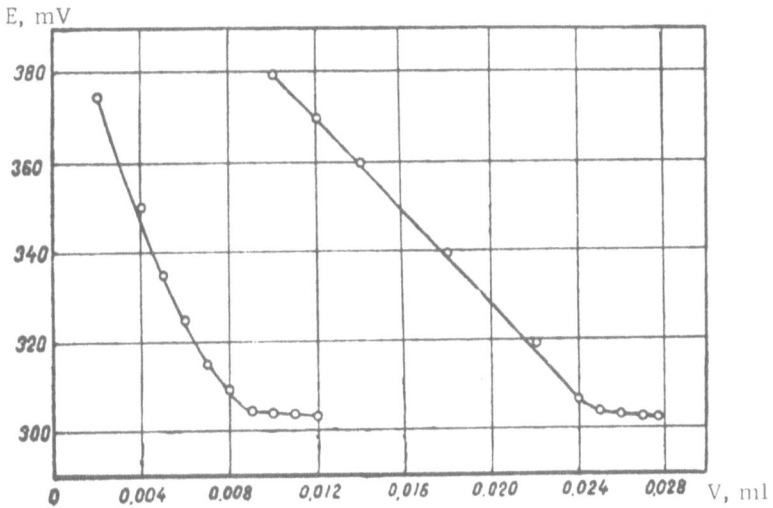

Fig. 1. Curves for the titration of indium using potentiometric determination of the end of titration: I) dependence of the potential on the amount of the added sodium ferrocyanide solution in the determination of pure indium; II) the same dependence but for indium in a ternary In−Sb−Te alloy.

TABLE 1

Sample weight, mg	Solvent volume, ml	Theoretical composition, in wt. %			Found, wt. %		
		In	Sb	Te	In	Sb	Te
0.36	10				47.7	12.2	39.8
0.20	10	47.69	12.65	39.76	48.3	14.7	37.5
0.34	5				44.2	14.5	40.4
Average values					47 ±2	14±1	39±1

We did not obtain such a jump, probably because the sample weights were three orders of magnitude smaller than those used in [7].

The titration process of indium is shown graphically in Fig. 1. The equivalent point was determined from the establishment of a constant potential. Such a form of the curves is definitely possible in the titration of complexes.

To separate the various phases of the In−Sb−Te system, an attempt was made to use the selective dissolution of the phases. However, we were unable to find a suitable selective solvent for the various phases. Therefore, we separated the phases of interest to us by mechanical means. To take samples of the separate phases, N. T. Savel'ev and S. D. Remenko constructed, using PMT-3 type microhardness meters, an attachment for taking microsamples of various phases. This attachment made it possible to position exactly a drill over the place from which we wanted a sample and to control quite accurately the depth of drilling. The drills were up to 0.1 mm in diameter. Microstructure analyses carried out before drilling at three mutually perpendicular sections of the sample showed that the phase distribution was approximately the same throughout the sample. This led us to the conclusion that phases occupying a large area should have had sufficient depth. Therefore, in taking microsamples, we drilled to a depth not greater than the diameter of the area occupied by a given phase (up to 0.2 mm), which guaranteed that only one phase was obtained by drilling. In some cases, it was necessary to drill the areas occupied by the same phase in various parts of the section in order to increase the amount taken for analysis. In

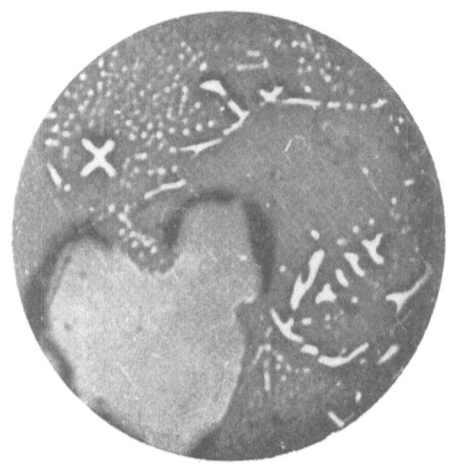

Fig. 2. Microstructure of the alloy $3In_3Sb_3 \cdot In_2Te_3$, which shows three phases; the two phases present in larger amounts were drilled for samples.

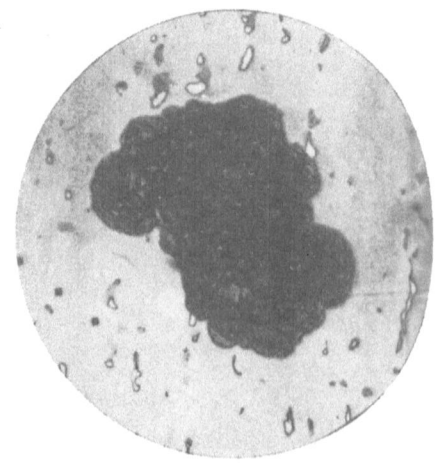

Fig. 3. Microstructure of the $In_3Sb_3 \cdot In_2Te_3$ alloy (front of the ingot) after zone leveling; the section is shown after drilling.

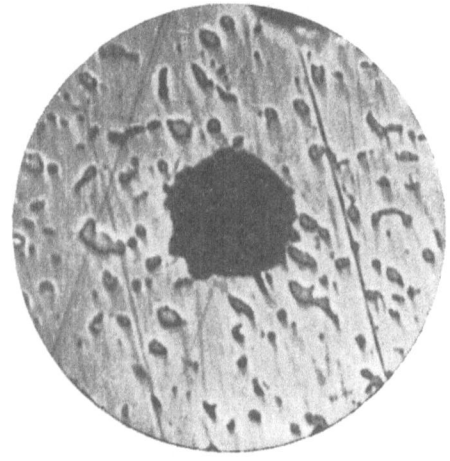

Fig. 4. Microstructure of the $In_3Sb_3 \cdot In_2Te_3$ alloy (middle of the ingot) after zone leveling.

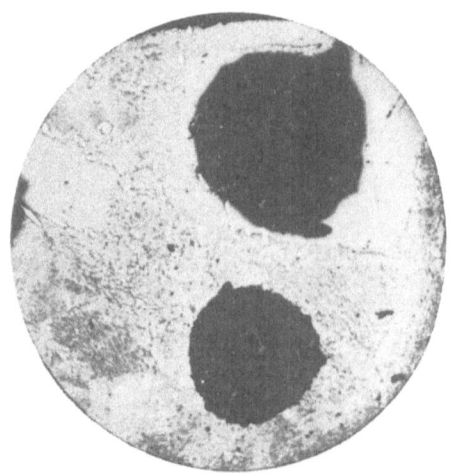

Fig. 5. Microstructure of the $In_3Sb_3 \cdot In_2Te_3$ alloy (end of the ingot) after zone leveling, showing drill holes for taking samples of the principal phase and mixed phase.

view of this, the problem of reproducibility of the chemical analysis data was of considerable importance. The uniformity of the phase was found by visual inspection using a metallurgical microscope (type MIM-7) and checked with a microhardness meter (type PMT-3). Moreover, the uniformity of the alloys was to be judged from x-ray diffraction data.

To check the accuracy of this microchemical analysis method, we drilled samples from a single-phase alloy represented by the formula In_4SbTe_3.

This alloy had, according to x-ray diffraction analysis, the rock-salt structure with a lattice parameter of 6.12_8 A. The microhardness was the same in various parts of a sample of this alloy and equal to 120 ± 5 kg/mm^2.

The microchemical analysis data for three samples of In_4SbTe_3, drilled at different locations in a section, are given in Table 1. They show that the error in the determination of the elements did not exceed 2 abs. %.

TABLE 2

Phases	Micro-hardness kg/mm²	Sample weight, mg	Solvent volume, ml	Found, wt. %		
				In	Sb	Te
Gray	230 ± 7	0.34	10	52.9	31.2	14.6
		0.49	10	49.0	37.9	12.7
		0.21	5	57.1	37.9	11.9
			Average	53 ± 3	35 ± 3	13 ± 1
Light-gray	70 ± 2	0.32	10	69.7	Not detected	31.2
		0.20	10	66.6		31.2
		0.21	10	66.6		30
			Average	67 ± 1	—	31 ± 0.5
Light	124 ± 4			Unable to isolate sample		

The chemical analysis data gave the molar composition of this compound as:

$$\text{In} : \text{Sb} : \text{Te} = \frac{46.0}{114.8} : \frac{14}{121.7} : \frac{39}{127.6} = 0.40 : 0.12 : 0.31 \approx 4 : 1 : 3,$$

which confirmed the composition represented by In_4SbTe_3.

Next, we investigated an alloy of composition $3In_3Sb_3 \cdot In_2Te_3$. Figure 2 shows a photomicrograph of this alloy in which three phases can be distinguished: light, light-gray, and gray. These phases differed in micro-hardness (Table 2). X-ray structure analysis of this alloy confirmed the presence of three phases.

The principal phase of the alloy had the zinc blende structure with a lattice parameter of 6.44 A; the second phase was of low symmetry, and the third had the rock-salt structure.

It was possible to carry out the microchemical analysis only for the two phases present in larger amounts: the gray and light-gray phases. Table 2 lists the results of this analysis (repeated three times for each phase) for the alloy $3In_3Sb_3 \cdot In_2Te_3$.

The results (Table 2) indicated that the gray phase contained all three elements. Their ratios, the value of the microhardness, and the lattice parameters indicated that this phase was a solid solution based on InSb.

Analysis of the light-gray phase showed only the presence of indium and tellurium; the ratio of these elements was close to that in In_2Te.

Thus, the microchemical analysis showed that the alloy $3In_3Sb_3 \cdot In_2Te_3$ did not contain the initial compounds InSb and In_2Te_3.

The microchemical phase analysis was used also to study an ingot, produced by zone leveling, of an alloy with the composition $In_3Sb_3 \cdot In_2Te_3$. The microstructure analysis of the end and middle regions showed that the last zone passage produced at the front of the ingot (Fig. 3) an alloy containing about 3% of the light phase on a general background of the gray phase. In the middle of the ingot (Fig. 4), the light phase content was somewhat greater. The value of the microhardness of the gray phase was equal to that of the ternary compound In_4SbTe_3.

TABLE 3

	Phases and their % composition	Micro-hardness, kg/mm²	Lattice type	Lattice parameter, Å	Description of sample	Sample weight, mg	Solvent volume, ml	Found, wt. %		
								In	Sb	Te
Front of ingot	gray, 97 · ·	120±4	NaCl	6.12₅	Mixture of 2 phases	0.77	10	45.4	20.6	34.1
	light, 3 · ·					0.23	10	43.5	23.1	32.6
					Average			44 ±1	22±1	33±1
Middle of ingot	gray, 95 · ·	120±5	NaCl	6.12₅	Mixture of 2 phases	0.27	10	25.9	39.5	37.0
	light, 5 · ·		ZnS	6.44		0.26	10	23	40.7	33.6
					Average			24 ±1	40±1	35±2
End of ingot	phase I. 80 · ·	−130±8	ZnS	6.47	Single phase	0.13	5	51.3	50.9	
						0.26	10	52.8	50.9	
	phase II. 20 · ·	−115±7	NaCl	6.13	One-phase and two-phase mixture	0.13	5	57.7	29.4	9.6
						0.25	10	54.8	31.8	15.0
					Average			56±2	31±1	12±2

It was not possible to carry out chemical analyses of the separate phases of the $In_3Sb_3 \cdot In_2Te_3$ alloy since the light phase (as shown in the photomicrographs) was present in too small amounts and could not be separated mechanically. Therefore, the chemical analysis reflected some average composition of the sample.

At the end of the ingot (Fig. 5) two phases were seen. The principal phase [1] occupied about 80% of the section area and had a microhardness of the order of 230 kg/mm^2. The second phase [1] was present in the form of small occlusions in the principal phase. The microhardness of the second phase was the same as that of the gray phase of the first two sections shown in Figs. 3 and 4.

The x-ray structure analysis of the front of the ingot showed the presence of the rock-salt structure with a lattice constant of 6.12$_5$ A. This phase also predominated in the middle part of the ingot, where the second phase with the zinc blende structure and a lattice parameter 6.44 A was also present. The end of the ingot consisted, according to the x-ray structure analysis, of two phases with the structures mentioned above, but the lattice parameter of the zinc blende structure was different: 6.47 A. Samples of the principal phase and of a mixture of the two phases were taken from this part of the ingot.

The microchemical analysis showed (Table 3) that the ratio of the elements at the front of the ingot was close to the ratio in the ternary compound In_4SbTe_3. The somewhat higher content of indium and antimony led to the appearance of the second phase. Analysis of the middle of the ingot showed a reduction of the indium content and an increase of the antimony content. The end of the ingot consisted mainly, according to the chemical analysis, of the phase whose chemical composition corresponded to InSb and a second phase corresponding to In_4SbTe_3. These chemical analysis data were in agreement with the microstructure and x-ray structure analyses.

CONCLUSIONS

1. A microchemical phase analysis method was developed for alloys in the In−Sb−Te system.

2. This method was used to investigate complex alloys. The results made it possible to determine the chemical composition of the phases and to study the changes in the composition on homogenization by zone leveling.

LITERATURE CITED

1. G. A. Kiosse, T. I. Malinovskii, and S. I. Radautsan, Izv. Moldavsk. Filiala Akad. Nauk SSSR, No. 3(69): 3 (1960).
2. S. I. Radautsan and I. P. Molodyan, Izv. Moldavsk. Filiala Akad. Nauk SSSR, No. 3(69): 37 (1960).
3. N. A. Goryunova, S. I. Radautsan, and G. A. Kiosse, Fiz. Tverd. Tela 1: 1858 (1959).
4. Yu. S. Lyalikov and L. S. Kopanskaya, Izv. Akad. Nauk Moldov.SSR, No. 12: 47 (1961).
5. Yu. S. Lyalikov and N. N. Safronkova, Zavodsk. Lab. 27: 21 (1961).
6. Yu. S. Lyalikov, L. S. Kopanskaya, and N. N. Safronkova, Proceedings of the Twentieth Scientific Conference of the Leningrad Building Institute, Physics Section, Leningrad (1962), p. 26.
7. É. N. Deichman and I. V. Tananaev, Zh. Analit. Khim. 13: 2 (1958).
8. I. V. Tananaev and M. I. Levina, Zavodsk. Lab. 15(8) (1949).

SOLID SOLUTIONS BASED ON INDIUM ANTIMONIDE
IN THE INDIUM – ANTIMONY – TELLURIUM SYSTEM

I. P. Molodyan and S. I. Radautsan

In recent years, the amount of work on solid solutions based on the most promising binary diamond-like semiconductors (such as indium antimonide, gallium arsenide, indium arsenide, etc.) has greatly increased. Separate tie-lines (in most cases, pseudobinary tie-lines) of various ternary systems have usually been investigated.

A study was made in our laboratory in 1959 of two tie-lines of the ternary system indium–antimony–tellurium, namely, $(InSb)_{3x} (In_2Te_3)_{1-x}$ and $(InSb)_x (InTe)_{1-x}$.

The properties of binary compounds of this system were reported in [1-5].

X-ray structure and microstructure studies showed that solid solutions are formed along the $(InSb)_{3x} (In_2Te_3)_{1-x}$ tie-line over a narrow range of concentrations near InSb [5]. Various methods of homogenization have not greatly extended the solubility range.

A limited homogeneous region near InSb and a new ternary compound with the rock-salt structure and a nominal formula In_4SbTe_3 were discovered along the $(InSb)_x (InTe)_{1-x}$ section [6]. Some physicochemical properties of In_4SbTe_3 have been investigated too [7].

Woolley et al. [8] also discovered solid solutions along the $(InSb)_{3x} (In_2Te_3)_{1-x}$ tie-line in the 15 equimol. % range and confirmed the existence of a new phase with the rock-salt structure. They investigated as well some electrical properties in the solid solution region.

Greenaway and Cardona [9] measured the reflectivity of these alloys to determine their energy band structure. The peaks in the dependence of the reflectivity on energy were explained by them as being due to interband transitions along the crystallographic directions [111] and 100].

Rosenberg and Strauss [10] investigated the tie-line $In_2Te_3 - Sb_2Te_3$ and have found that 45 mol. % In_2Te_3 dissolves in Sb_2Te_3 without altering the crystal lattice of the latter (rhombohedral structure).

It is evident from these results that the ternary system indium–antimony–tellurium has not been fully investigated.

Bearing in mind the suggestion of Goryunova [11] that the formation of solid solutions is possible not only along certain tie-lines of the system but also in certain areas near the binary compounds, it is interesting to investigate the possibility of the appearance of homogeneous regions near InSb in the ternary system indium –antimony–tellurium.

The present paper reports the results of physico-chemical and electrical studies of alloys of the $(InSb)_x \cdot (InTe)_{1-x}$ tie-line, and preliminary data on the formation of solid solutions along the $(InSb)_{2x} (In_2Te)_{1-x}$, $(InSb)_{5x} (In_2Te_5)_{1-x}$, and $(InSb)_{7x} (In_4Te_7)_{1-x}$ tie-lines in the ternary system indium–antimony–tellurium.

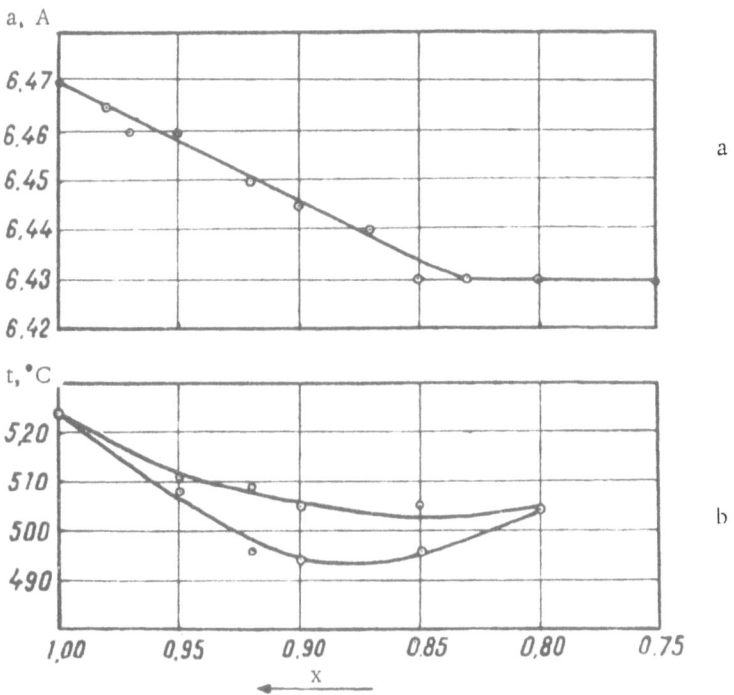

Fig. 1. Dependence of the lattice parameter on the composition (a) and the solidus and liquidus lines (b) for alloys along the tie-line $(InSb)_x (InTe)_{1-x}$.

SAMPLE PREPARATION AND MEASUREMENT METHODS

The samples were prepared from the component elements of not less than 99.999% purity. The alloys were made by direct fusion in quartz ampoules in an argon atmosphere using vibration mixing [12]. The samples were polycrystalline. Microchemical analysis of the homogeneous alloys showed that their chemical composition was the same as in the initial charge.

The alloys were investigated by x-ray diffraction, thermal, and microstructure methods. The temperature dependences of the electrical conductivity, the Hall coefficient, and the thermoelectric power were determined. The x-ray diffraction analysis was carried out using the Debye—Scherrer method in RKU-114 and RPK-2 cameras with copper radiation and nickel filters. A metallurgical microscope of the MIM-7 type was used for the microstructure analysis. The microhardness of the same samples was measured with a PMT-3 type instrument. The thermal analysis was carried out using a Kurnakov pyrometer (type FPK-55). A compensation circuit was employed for the electrical measurements. The measurements were carried out in a vacuum or in an argon atmosphere between room temperature and 350°C. The sample-holder described in [13] was used. Tungsten probes of 0.2 mm diameter were soldered to samples measuring $10 \times 3 \times 1.5$ mm. The thermoelectric power was measured with respect to the copper branches of copper—constantan thermocouples, using a temperature drop of 10-15 deg between the sample ends. The Hall effect was measured in a magnetic field of 10^4 Oe, using a current of 100 mA.

RESULTS OF THE INVESTIGATION

Table 1 lists the compositions and some physico-chemical properties of several of the investigated alloys of the $(InSb)_x (InTe)_{1-x}$ tie-line.

The Debye diffraction patterns of the alloys having the compositions x = 1-0.85 showed clearly only one system of lines, characteristic of the zinc blende structure. The lattice parameter decreased linearly from 6.47 A for x = 1 to 6.43 A for x = 0.85 (Fig. 1a). The microstructure analysis showed that the alloys having the compositions x = 1-0.85 consisted of a single phase.

TABLE 1

Composition	Number of phases	Percentage ratios of phases	Lattice type	Lattice parameter, A	Micro-hardness
1	one	100	ZnS	6.47	217
0.98	"	100	ZnS	6.46_5	219
0.95	"	100	ZnS	6.46	228
0.90	"	100	ZnS	6.44_5	229
0.85	"	100	ZnS	643	230
0.83	two	95	ZnS	6.43	235
		5	NaCl	—	119
0.80	"	92	ZnS	6.43	237
		8	NaCl	—	126
0.50	"	35	ZnS	6.42_8	236
		65	NaCl	6.12_5	128
0.30	"	10	ZnS	6.43	235
		90	NaCl	6.12_6	130
0.27	"	3	ZnS	—	233
		97	NaCl	6.12_5	122
0.25	one	100	NaCl	6.12_8	124
0.23	two	3	low-symmetry	—	86
		97	NaCl	6.12_5	113
0.10	two	65	low-symmetry		87
		35	NaCl	6.12_7	115
0.00	one	100	low-symmetry	—	89

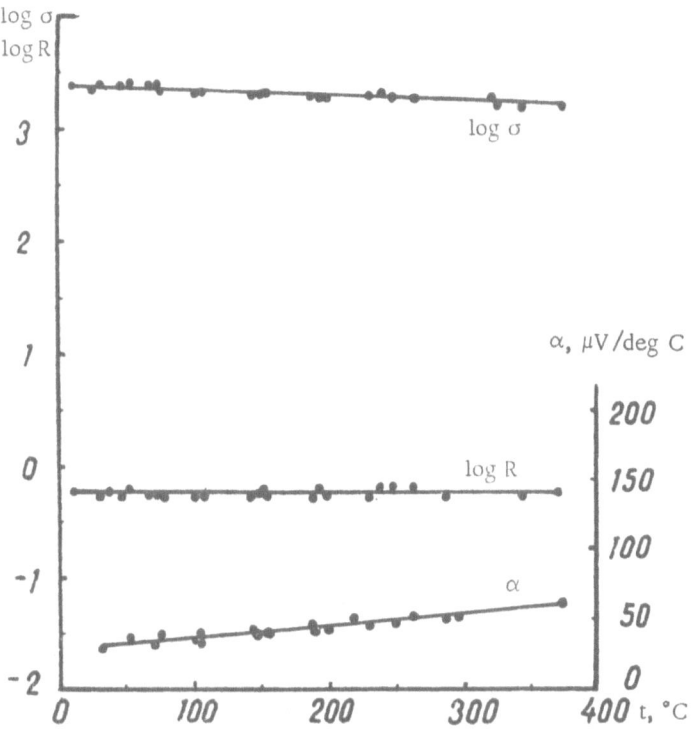

Fig. 2. Temperature dependences of log σ, log R, and α for an alloy having the composition $(InSb)_{0.97}(InTe)_{0.03}$.

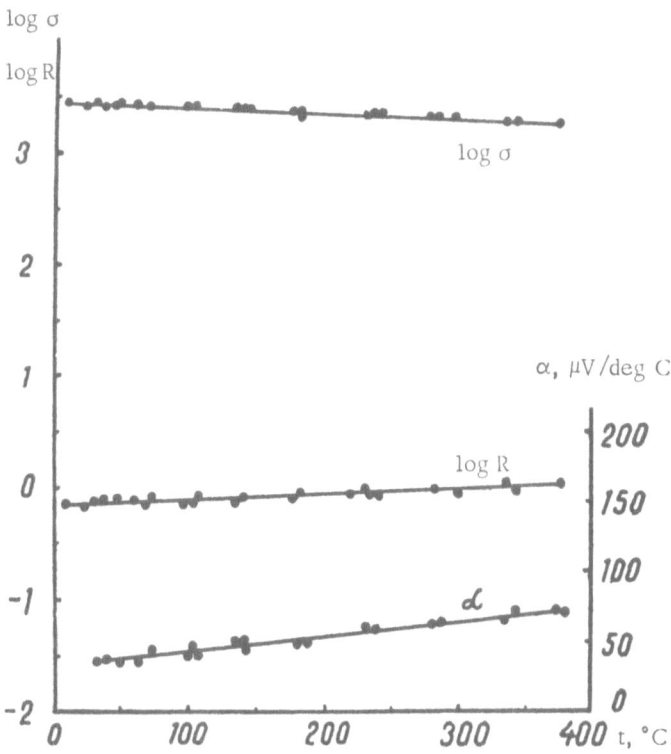

Fig. 3. Temperature dependences of log σ, log R, and α for an alloy having the composition $(InSb)_{0.9}(InTe)_{0.1}$.

The thermal analysis showed clearly the point of solidification. The solidus and liquidus curves for this region are illustrated in Fig. 1b.

The reported results indicated that solid solutions were formed along the tie-lines $(InSb)_x (InTe)_{1-x}$ at $x = 1-0.85$.

The Debye patterns of the alloys with the compositions $0.25 < x < 0.85$ exhibited, apart from the lines of the ZnS structure, lines representing the rock-salt structure. The intensity of the ZnS lines decreased but the intensity of the NaCl lines increased with increasing x. The Debye pattern of the alloy having $x = 0.25$ exhibited only one system of lines corresponding to the NaCl type structure with the lattice parameter 6.12_8 A. This phase represented the ternary compound In_4SbTe_3 [6]. On further decrease in the InSb content ($x < 0.25$) the Debye patterns showed low-symmetry lines characteristic of the compound InTe. Microstructure analysis showed that the alloys in the composition range $0.25 < x < 0.85$ consisted of two phases whose microhardness indicated that they corresponded to the compostions $x = 0.85$ and $x = 0.25$. A new phase appeared at $x < 0.25$, having the microhardness characteristic of InTe, while the phase with the composition $x = 0.25$ was retained.

The presence of two phases in the alloys with $0.25 < x < 0.85$ and $0 < x < 0.25$ was also confirmed by thermal analysis.

We measured the temperature dependences of the electrical conductivity, the thermoelectric power, and the Hall coefficient for the alloys having the compositions $x = 1-0.8$. These measurements showed that in this region all the alloys were n-type. The dependences of the Hall coefficient on the magnetic field intensity (up to 10^4 Oe), recorded for samples with $x = 0.97$, 0.9, and 0.85, showed that in the range of investigated temperatures R was independent of the field.

Figures 2 and 3 show the temperature dependences of the electrical conductivity, the Hall coefficient, and the thermoelectric power for the two compositions $x = 0.97$ and $x = 0.90$. Similar dependences were obtained for other alloys in the solid solution range. The conductivity and the Hall coefficient depended weakly on temperature and, therefore, the temperature dependence of the density and the carrier mobility was weak. These dependences indicated election-gas degeneracy in the alloys.

Figure 4 shows the dependences of the electrical conductivity, the Hall coefficient, and the thermoelectric power on the alloy compostion. The electrical conductivity rose sharply while the thermoelectric power and Hall coefficient decreased on addition of 0.1% InTe to InSb. The dependences of the carrier density and mobility on the alloy composition, shown in Fig. 5, indicated that the electron density rose sharply from 7×10^{16} cm^{-3} for the alloy with $x = 1$ to 10^{19} cm^{-3} for the alloy with $x = 0.999$, but on further reduction in the value of x, the change was small. The mobility decreased from 48,500 cm$^2 \cdot$ V$^{-1} \cdot$ sec^{-1} for InSb to 1500 cm$^2 \cdot$ V$^{-1} \cdot$ sec^{-1} for the alloys with $x = 0.97$, but altered little on further reduction in x.

DISCUSSION OF THE RESULTS

The quoted results of the x-ray diffraction, microstructure, and thermal analyses show that solid solutions with the zinc blende structure are formed along the $(InSb)_x (InTe)_{1-x}$ tie-line in the concentration region $x = 1-0.85$. The lattice parameter decreases linearly from 6.47 A for $x = 1$ to 6.43 A for $x = 0.85$. It is known [8] that solid solutions are formed also along the $(InSb)_{3x} (In_2Te_3)_{1-x}$ tie-line. It is interesting that along these two tie-lines of the ternary system In−Sb−Te there are identical solubility regions with the same change in the lattice parameter. To obtain information on the solubility in indium antimonide and other compounds and phases of the binary system In−Te, we prepared and investigated, by x-ray diffraction and microstructure methods, four samples for each of the compositions $x = 0.95$, 0.90, 0.85, and 0.80 along the following tie-lines: $(InSb)_{2x} (In_2Te)_{1-x}$, $(InSb)_{5x} (In_2Te_5)_{1-x}$, and $(InSb)_{7x} (In_4Te_7)_{1-x}$.

The results show that in the region $x = 1-0.85$ the alloys along all these tie-lines consist of a single phase with the zinc blende structure; the parameter of this structure varies according to Vegard's law from 6.47 A for $x = 1$ to 6.43-6.44 A for $x = 0.85$. The alloys with the composition $x = 0.8$ consist of two phases, according to the results of the x-ray diffraction and microstructure analyses. Thus, in the ternary system In−Sb−Te in the region of 15 equimol. % near InSb, solid solutions InSb compound are formed with all other compounds and phases

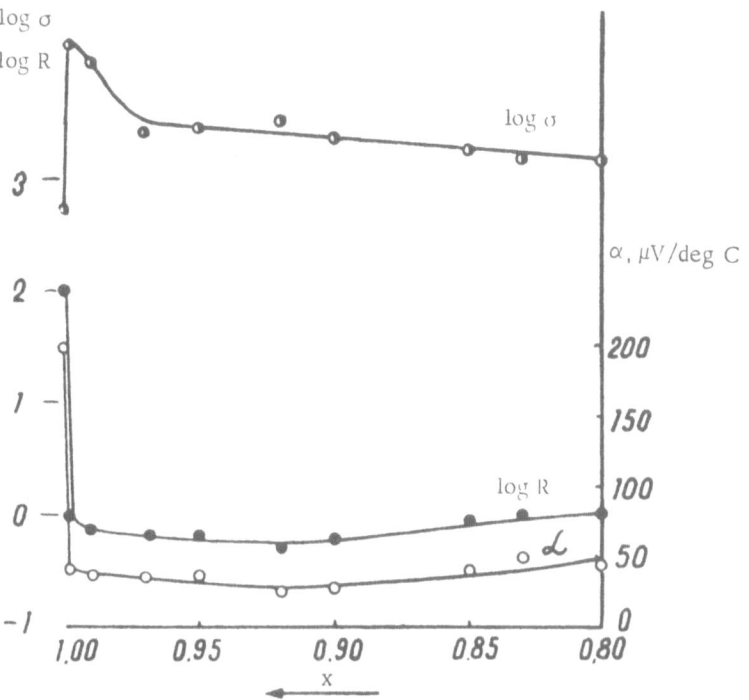

Fig. 4. Dependences of log σ, log R, and α on the composition of alloys along the tie-line $(InSb)_x (InTe)_{1-x}$.

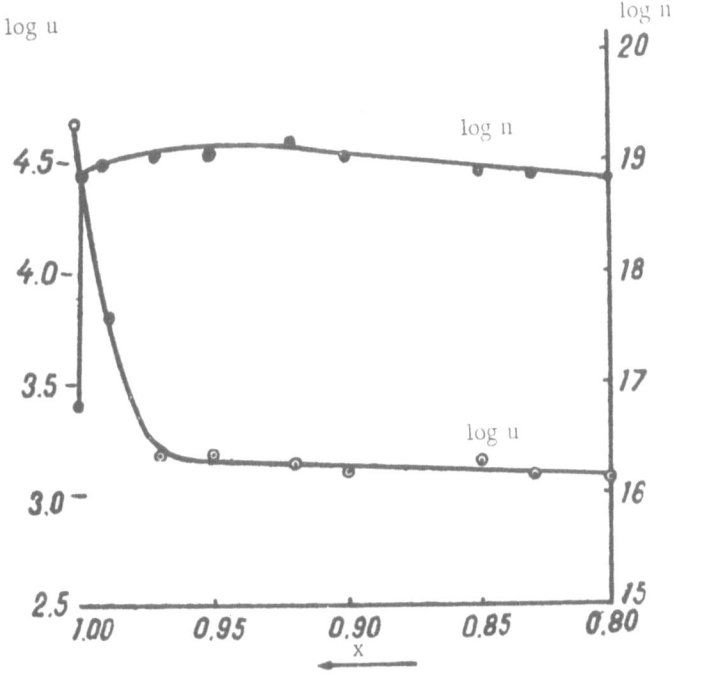

Fig. 5. Dependences of log n and log u on the composition of alloys along the tie-line $(InSb)_x (InTe)_{1-x}$.

of the binary system $In-Te$. These solid solutions have the zinc blende structure with the lattice parameters varying within the same limits for all three tie-lines. The solid-solution structures corresponding to different tie-lines of the $In-Sb-Te$ system differ in the amount and distribution of the defects in the anion or cation sublattice, depending on the ratio of elements in the indium telluride system.

Any similarity of the solid solution structures corresponding to different sections of the system $In-Sb-Te$ is complemented by the similarity of their electrical properties. Thus, the electrical conductivity and the Hall coefficient of the solid solutions are practically independent of temperature, both for the tie-line $(InSb)_{3x} \cdot (In_2Te_3)_{1-x}$ and for the tie-line $(InSb)_x (InTe)_{1-x}$. Small amounts of InTe, like In_2Te_3, sharply increase the carrier density. It is known [14, 15] that such a dependence of the carrier density on composition with a maximum near $A^{III}B^V$ compounds is characteristic of solid solutions formed by the interaction of $A_2^{III}B_3^{VI}$. Our results indicate that a similar dependence of the density on the alloy composition applies also in the case of solid solutions along the tie-line $(InSb)_x (InTe)_{1-x}$, where both these compounds are of the nondefect type.

CONCLUSIONS

1. Solid solutions were discovered along the tie-line $(InSb)_x (InTe)_{1-x}$ in the concentration region $x = 1-0.85$.

2. The temperature dependences of the electrical conductivity, the Hall coefficient and the thermoelectric power were investigated for polycrystalline alloys of the tie-line $(InSb)_x (InTe)_{1-x}$ in the region $x = 1-0.8$.

3. The electrical conductivity maximum was found for the alloys having the composition $x = 0.999$, in the region close to InSb.

The electrical and galvanomagnetic properties depended strongly on the composition only near InSb; in alloys containing large amounts of InTe, these properties depended weakly on the composition and temperature.

4. Solid solutions were also discovered along the tie-lines $(InSb)_{2x} (In_2Te)_{1-x}$, $(InSb)_{5x} (In_2Te_5)_{1-x}$, and $(InSb)_{7x} (In_4Te_7)_{1-x}$ in the concentration region $x = 1-0.85$, which confirms experimentally the hypothesis of the existence of wide homogeneous regions in alloys based on indium antimonide.

LITERATURE CITED

1. M. Hansen and K. Anderko, Structure of Binary Alloys, Metallurgizdat, Moscow (1962).
2. S. Sugaike and S. M. Takabayasi, Japanese Patent, No. 2076 (1955).
3. Wright and Brice, Nature 183: 33 (1959).
4. N. I. Volokobinskaya, V. V. Galavanov, and D. N. Nasledov, Fiz. Tverd. Tela 1: 755 (1959).
5. S. I. Radautsan and I. P. Molodyan, Izv. Moldavsk. Filiala Akad. Nauk SSSR, No. 3(69): 37 (1960).
6. N. A. Goryunova, S. I. Radautsan, and G. A. Kiosse, Fiz. Tverd. Tela 1: 1858 (1959).
7. I. P. Molodyan, S. I. Radautsan, and I. A. Madan, Izv. Akad. Nauk Moldav.SSR, No. 10: 88, 57 (1961).
8. J. C. Woolley, C. M. Gillett, and J. A. Evans, J. Phys. Chem. Solids 16: 138 (1960).
9. D. L. Greenaway and M. Cardona, Proceedings of the International Conference on the Physics of Semiconductors, Exeter (1962).
10. A. J. Rosenberg and A. J. Strauss, J. Phys. Chem. Solids 19: 105 (1961).
11. N. A. Goryunova, Paper presented at the All-Union Conference on Semiconducting Compounds, Leningrad (1961).
12. A. S. Borshchevskii and D. N. Tret'yakov, Fiz. Tverd. Tela 1: 1483 (1959).
13. O. V. Emel'yanenko and N. V. Trishin, PTE, No. 6: 130 (1959).
14. D. N. Nasledov and I. A. Feltyn'sh, Fiz. Tverd. Tela 1: 565 (1959); 2: 823 (1960).
15. J. C. Woolley and P. N. Keating, Proc. Phys. Soc. (London) 78: 503, 1009 (1961).

SOME COMPLEX PHASES BASED ON In_2Te_3

S. I. Radautsan, R. A. Maslyanko, and M. M. Markus

Investigations of multicomponent systems have shown that it is possible to form ternary, quaternary, and more complex semiconducting compounds which are analogs of the binary compounds with the zinc blende structure [1, 2].

In some cases, wide solubility regions have been discovered in which there are interactions between these complex compounds [3-5].

The present paper reports the preliminary results of a study of tie-lines which, to the authors' knowledge, have not yet been investigated: $(AgInTe_2)_{3X} - (In_2Te_3)_{2(1-x)}$ of the ternary system silver—indium—tellurium, as well as $(CdIn_2Se_4)_X - (CdIn_2Te_4)_{1-x}$ and $(Cd_2SeTe)_{3X} - (In_4Se_3Te_3)_{1-x}$ of the quaternary system cadmium—indium—selenium—tellurium. These particular tie-lines were selected because some materials with a valuable combination of properties have been prepared earlier using defect compounds of the $A_2^{III}B_3^{VI}$ type [6-8]. Zone melting of one of the initial phases — $CdIn_2Te_4$ — produced a high value of the mobility ($4000 \ cm^2 \cdot V^{-1} \cdot sec^{-1}$) at an electron density of $10^{14} \ cm^{-3}$ [9] in spite of a considerable number of neutral "intrinsic" defects in the lattice.

Alloys prepared by the standard method [10] were investigated by means of x-ray diffraction and microstructure analyses and their microhardness was measured. Two or three samples were prepared for each composition. The x-ray diffraction work was carried out using camera types RKU-86 and RKD-57, employing copper and iron radiation; microsections were examined with metallurgical microscope type MIM-7, and the microhardness was measured with a PMT-3-type instrument.

$(CdIn_2Se_4)_X - (CdIn_2Te_4)_{1-x}$ TIE-LINE

Alloys having the compositions $x = 0, 0.1, 0.2, 0.25, 0.3, 0.4, 0.5, 0.6, 0.7, 0.75, 0.9, 1$ were prepared. The samples were of moderate density and gray in color. The Debye powder patterns of all compositions in the range $x = 0-0.9$ exhibited the lines of a thiogallate structure of the $CdIn_2Te_4$ type (S_4^2 group), investigated earlier by Hahn [11]. They differed from the Debye pattern of $CdIn_2Se_4$ (D_{2d}^1 group). The lattice constant "a" varied continuously from 6.23 A for $x = 0$ to 5.82 A for $x = 1$ (Fig. 1).

Microstructure investigations showed that all alloys consisted mainly of a single phase. Their microhardness varied from 220 kg/mm^2 for $CdIn_2Te_4$ to 300 kg/mm^2 for $CdIn_2Se_4$. For alloys of the composition $x = 0.5$, there was a large scatter of the microhardness values, probably due to the insufficient homogenization of these samples.

Thus, along the tie-line $(CdIn_2Se_4)_X - (CdIn_2Te_4)_{1-x}$ solid solutions were obtained over the whole range of concentrations, and the structure of $CdIn_2Te_4$ type was retained at least up to $x = 0.9$.

$(Cd_2SeTe)_{3X} - (In_4Se_3Te_3)_{1-x}$ TIE-LINE

To find whether solid solutions could be formed in the quaternary system cadmium—indium—selenium—tellurium along other tie-lines, we prepared three compositions for the tie-line $(Cd_2SeTe)_{3X} - (In_4Se_3Te_3)_{1-x}$ for $x = 0.75, 0.25$, and 0.1. The Debye patterns of all the alloys exhibited the zinc blende structure of the sphalerite

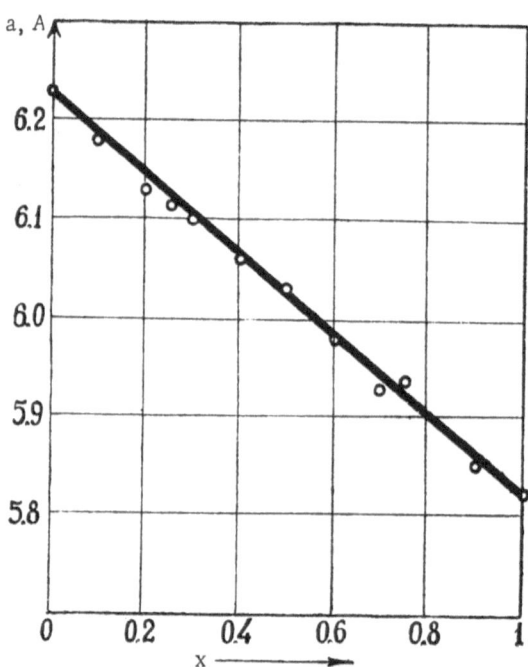

Fig. 1. Dependence of the lattice parameter on the composition of alloys of the system $(CdIn_2Se_4)_x$ $-(CdIn_2Te_4)_{1-x}$

TABLE 1. Lattice Parameters of Alloys Along the Tie-Line $(Cd_2SeTe)_{3x}-(In_4Se_3Te_3)_{1-x}$

Composition, x =	0.75	0.5	0.25	0.1
Structure type	sphalerite	thiogallate	sphalerite	sphalerite
Lattice parameter, in A	6.20	6.03	6.03	5.97

TABLE 2. Dependence of Lattice Parameters on Concentration Along the Tie-Line $(AgInTe_2)_{3x}-(In_2Te_3)_{2(1-x)}$.

Composition, x =	0	0.1	0.2	0.25	0.3	0.4	0.5	0.6	0.7	0.8	0.9	1
Structure	ZnS	thiogallate						two phases	chalcopyrite			
"a," in A	6.16	6.17	6.19	6.20	6.22	6.24	6.26		6.36	6.39	6.40	6.43
"c," in A	—	12.34	12.38	12.40	12.44	12.48	12.52		12.50	12.56	12.58	12.64
c/a	—	2	2	2	2	2	2		1.97	1.97	1.97	1.97

modification. Table 1 shows that along this tie-line a monotonic variation of the lattice parameter was again observed. Only the composition with x = 0.5 (Fig. 1) differed from the other alloys; ordering (thiogallate structure) and a smaller lattice parameter were observed for this composition.

In summarizing this preliminary study of two tie-lines of the quaternary system cadmium—indium—selenium—tellurium, we may conclude that wide solubility regions occur in alloys of this system.

$(AgInTe_2)_{3x}-(In_2Te_3)_{2(1-x)}$ TIE-LINE

We prepared and investigated alloys whose compositions are listed in Table 2. The samples were gray in color with a metallic luster. The compound In_2Te_3 showed superstructure lines in its Debye patterns, which indicated the ordering of one-third of the cation vacancies [12]. The x-ray diffraction analysis showed that the alloys having compositions $x = 0.1-0.5$ had the thiogallate structure, similar to the structure of $CdIn_2Te_4$. The lines distinguishing this structure from zinc blende were strongest in the Debye patterns for the "stoichiometric" compositions $x = 0.25$ (nominal formula $AgIn_5Te_8$). In this phase, one-quarter of the cation vacancies was ordered compared with $CdIn_2Te_4$ and this ordering was retained both for larger and smaller numbers of intrinsic defects. Both lattice parameters increased with the concentration of silver in the alloys (Table 2). The Debye patterns of the alloys with $x = 0.7-1$ had only the chalcopyrite structure lines; both lattice parameters increased. Lines indicating two phases were observed at $x = 0.6$, corresponding to the structures of chalcopyrite and thiogallate.

The microstructure analysis revealed that alloys with the compositions $x = 0-0.5$ consisted of a single phase whose microhardness increased monotonically from 180 to 240 kg/mm². For compositions with the chalcopyrite structure a large scatter of the microhardness values was observed, which perhaps indicated insufficient homogeneity of the alloys.

From the results obtained, we conclude that along this tie-line there are two regions of solid solutions differing in structure. The samples of compositions close to In_2Te_3, a defect compound with the zinc blende structure, are more homogeneous and have larger grains.

Similar observations were reported by Rodot [13], who investigated systems based on $AgInTe_2$, including cadmium and mercury tellurides.

LITERATURE CITED

1. N. A. Goryunova, Vestn. Leningr. Gos. Univ. 10: 112 (1961).
2. C. H. L. Goodman, J. Phys. Chem. Solids 6: 305 (1958).
3. N. A. Goryunova and V. D. Prochukhan, Fiz. Tverd. Tela 2: 176 (1960).
4. N. A. Goryunova, A. V. Voitsekhovskii, and V. D. Prochukhan, Vestn. Leningr. Gos. Univ. 10: 156 (1961).
5. S. I. Radautsan and R. A. Ivanova, Izv. Akad. Nauk Moldav.SSR, No. 10(88): 64 (1961).
6. B. T. Kolomiets and A. A. Mal'kova, Fiz. Tverd. Tela, Sbornik 2: 32 (1959).
7. V. D. Porchukhan, Some Quaternary Semiconducting Phases (author's abstract of dissertation for candidate's degree), Moscow (1962).
8. N. A. Goryunova and S. I. Radautsan, this volume, p. 1.
9. D. R. Mason and D. F. O'Kane, Proc. Intern. Conf. Semicond., held in Prague, 1960 (publ. Prague 1961), p. 1026.
10. A. S. Borshchevskii and D. N. Tret'yakov, Fiz. Tverd. Tela 1: 1483 (1959).
11. H. Hahn, G. Frank, W. Klingler, A. D. Störger, and G. Störger, Z. Anorg. Allgem. Chem. 5-6: 279, 241 (1955).
12. A. I. Zaslavskii and V. M. Sergeeva, Fiz. Tverd. Tela 2: 2872 (1960).
13. H. Rodot, Proc. Intern. Conf. Semicond. Phys., held in Prague, 1960 (publ. Prague 1961), p. 1010.

SOLID SOLUTIONS OF GALLIUM PHOSPHIDE AND SULFIDE

S. I. Radautsan and V. V. Negreskul

Semiconductor technology needs new effective materials suitable for high temperatures. Investigations show that some of the $A^{III}B^V$ compounds and solid solutions based on them satisfy the requirements. A special position is held among them by the heterovalent solid solutions formed between the compounds $A^{III}B^V$ and the defect compounds $A_2^{III}B_3^{VI}$. The first experimental results on the possibility of mutual solubility in such systems were obtained by Goryunova in 1955 [1].

The presence of solid solutions in a wide range of concentrations has been established for the system $GaAs-Ga_2Se_3$ [2]. Other systems of this type were studied later [3-5], and solutions have now been observed over the whole range of concentrations in the system $InAs-In_2Te_3$ [3, 4].

The electrical properties of gallium arsenoselenides [6], indium arsenotellurides [4], and alloys of other systems have shown that solid solutions with high $A^{III}B^V$ concentrations exhibit certain anomalies in the form of a sharp rise in the electrical conductivity and a marked increase in the carrier density. These phenomena may be associated with a complex mechanism of solution in the $A^{III}B^V$ compounds, which gives rise to wide solid solution ranges even outside the pseudobinary tie-lines of ternary systems [7].

The combination of a high electrical conductivity and a quite low thermal conductivity [8] suggests the application of some of these alloys in thermoelectric semiconducting devices.

Limited solubility has been observed in several solid solutions of the $A^{III}B^V-A_2^{III}B_3^{VI}$ type, for example, in the systems $InP-In_2Se_3$ [9], $InAs-In_2Se_3$ [3], $GaAs-Ga_2S_3$ [11], $GaAs-Ga_2Te_3$ [12], $AlSb-Al_2Te_3$ [13], etc.

The experimental results show that, in systems of this type, the interaction is not always the same. The formation of solid solutions with a tetrahedral distribution of atoms in the lattice depends on the similarity of the structures of the compounds, on the type of chemical binding, on the relationship between the atomic radii, and on other factors. The presence of solid solutions confirms the similarity of the compounds $A_2^{III}B_3^{VI}$ and the compounds of the $A^{III}B^V$ and $A^{II}B^{VI}$ type, not only in structure but also in the nature of the electron interaction [14]. In order to analyze the variation of the interaction with the composition in systems of the $A^{III}B^V-A_2^{III}B_3^{VI}$ type it is necessary to study systems containing elements belonging to different periods of Mendeleev's table. Among substances of this type are alloys of the $GaP-Ga_2S_3$ systems.

Recent investigations have shown that gallium phosphide may be used to make high-temperature thermocouples, solar batteries, and other devices [15]. Gallium sulfide has high photosensitivity and may be used in photoelectric devices. Some properties of these two compounds are listed in Table 1. In view of the good semiconducting properties of gallium phosphide and gallium sulfide, we selected the system $GaP-Ga_2S_3$ for investigation of the possibility of the formation of solid solutions which would combine the best properties of the initial compounds.

SAMPLE PREPARATION AND INVESTIGATION METHODS

We used high-purity (> 99.99%) elements to prepare alloys of the system $(GaP)_{3x}-(Ga_2S_3)_{1-x}$.

Samples were made by fusing the components in evacuated quartz ampoules. Vibration mixing [16] was used to accelerate the synthesis. The samples were obtained in the form of dense ingots.

TABLE 1. Some Properties of Gallium Phosphide and Sulfide

Binding	Structure type	Lattice parameter a, in A	Melting point, °C	Density, g/cm³	Microhardness, kg/mm²	ΔE_{opt} at 300 °K	Mobility, cm²·V⁻¹·sec⁻¹	
							electrons	holes
GaP	ZnS	5.45	~1500	4.15	940	2.24	110	75
Ga₂S₃	α, w**	a = 6.37 c = 18.05	1250	3.74	500	2.5	–	–
	β, w*	a = 3.678 c = 6.016						
	γ, ZnS	5.17						

Notation:

w* — defect wurtzite structure

w** — ditto, with ordered distribution of vacancies [11, 14, 18].

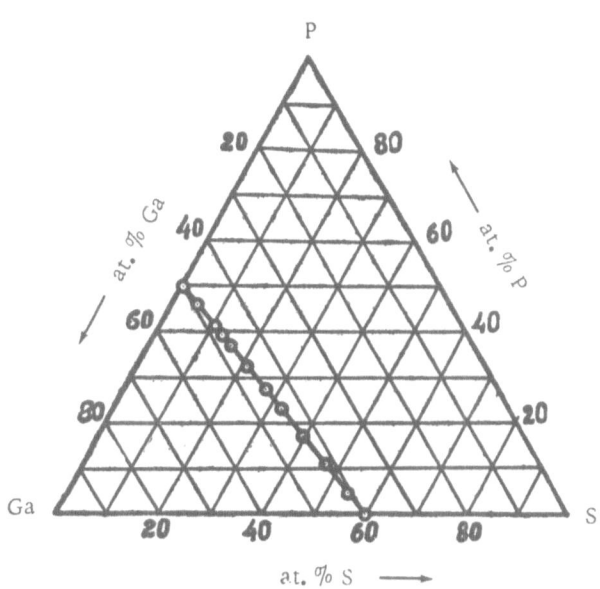

Fig. 1. Position of the investigated alloys in the pseudobinary system $(GaP)_{3x} - (Ga_2S_3)_{1-x}$.

The region of the existence of solid solutions was determined by x-ray structure and microstructure analyses. X-ray diffraction studies were carried out using a camera of the RKU-114 type, employing copper radiation and a nickel filter. The microhardness was measured with a PMT-3 type meter.

RESULTS OF THE INVESTIGATION

We investigated 12 alloys along the pseudobinary tie-line $(GaP)_{3x} - (Ga_2S_3)_{1-x}$, including the initial binary compounds. The compositions are shown in the gallium—phosphorus—sulfur triangle (Fig. 1).

X-ray diffraction studies showed that the addition of Ga₂S₃ to GaP produced phases with the zinc blende structure. The Debye patterns of all compositions down to x = 0.3 had the same system of strong lines in the concentration range of 70 equimol. % on the gallium phosphide side. The alloys with 0.3 > x > 0 had several phases

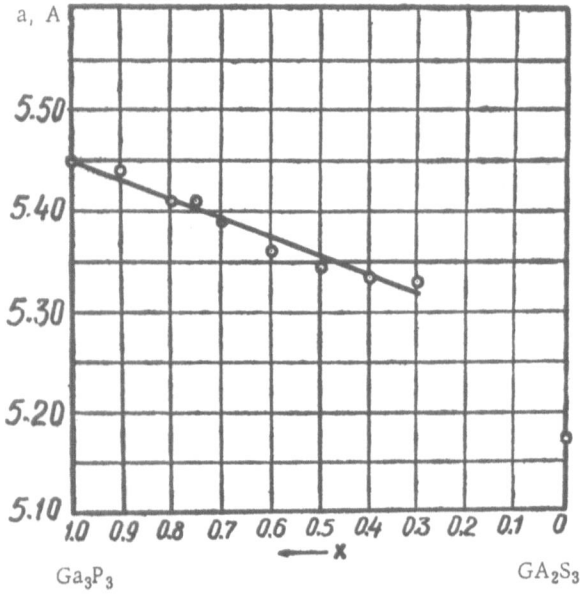

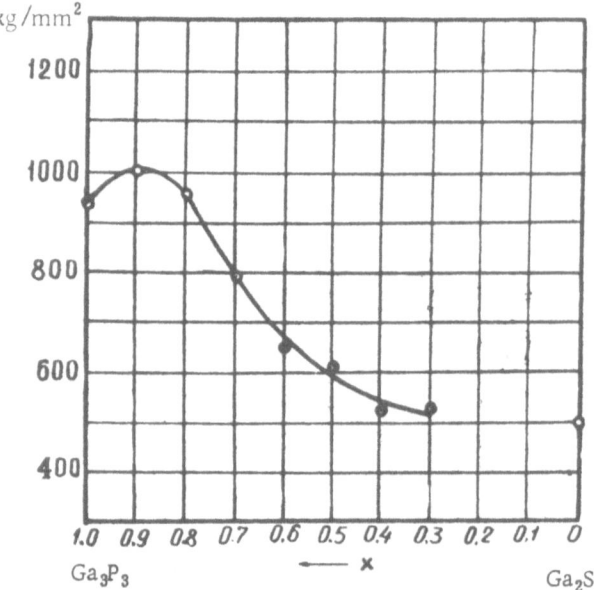

Fig. 2. Dependence of the lattice constant of alloys of the $(GaP)_{3x}-(Ga_2S_3)_{1-x}$ system.

Fig. 3. Dependence of the microhardness on the composition in the alloys of the system $(GaP)_{3x}-(Ga_2S_3)_{1-x}$.

and the Debye patterns showed additional lines which did not belong to the zinc blende structure. Thus, the results of the x-ray study showed the solubility of Ga_2S_3 in GaP in the range of concentrations from $x = 1.0$ to $x = 0.3$.

The reduction in the unit cell parameter from 5.45 A for gallium phosphide to 5.34 A for $x = 0.3$ was in good agreement with Vegard's law (Fig. 2) and indicated the formation of solid solutions.

The microstructure analysis of samples of this tie-line showed that they were uniform from gallium phosphide to $x = 0.3$. For the alloys in the concentration range $0.3 > x > 0$, several phases with different values of the microhardness were observed.

The results of the microhardness measurements are shown graphically in Fig. 3. The dependence of the microhardness on the composition had a maximum, observed earlier for similar systems [3, 4, 10].

All the alloys were in the form of dense polycrystalline ingots, whose color varied with composition. Gallium phosphide was transparent and orange. On the addition of Ga_2S_3 to GaP in the concentration range $1.0 > x > 0.3$, the alloys became dark orange. Red, yellow, and light yellow phases were observed in the nonuniform samples. Ga_2S_3 was light yellow in color, due to a small amount of bright yellow GaS present as an impurity [16].

Since the maximum temperature during the preparation of the alloys with $x = 0.7-0.5$ did not exceed 1200°C, which was lower than the melting points of the initial compounds (Table 1), the phase diagram for the gallium phosphide—sulfide system should be similar to those found earlier for analogus systems [10, 14].

It is quite likely that a selection of the optimum synthesis conditions and the application of homogenization methods may extend the solid-solution region of the investigated system.

Preliminary investigation of the electrical properties of the solid solutions $(GaP)_{3x}-(Ga_2S_3)_{1-x}$ showed that all the alloys had high resistivity and were semiconductors. Further detailed studies of their electrical, thermal, galvanomagnetic, and optical properties should allow us to determine the principal characteristics of these solid solutions.

CONCLUSIONS

X-ray structure and microstructure studies of the alloys along the pseudobinary tie-line $(GaP)_{3x}-(Ga_2S_3)_{1-x}$ of the ternary system gallium—phosphorus—sulfur showed the existence of GaP-based solid solutions with the zinc blende structure in the concentration range 70 equimol.%.

The lattice constant "a" varied linearly from 5.45 A for x = 1.0 to 5.34 A for x = 0.3, in good agreement with Vegard's law.

The solid solutions of these alloys were obtained free of additional phases directly after synthesis and they exhibited semiconducting properties.

LITERATURE CITED

1. N. A. Goryunova, in Collection: Problems of the Theory and Experimental Studies of Semiconductors and Semiconductor Metallurgy Processes, Izd. Akad. Nauk SSSR, Moscow (1955), p.29.
2. N. A. Goryunova and V. S. Grigor'eva, Zh. Tekhn. Fiz. 26: 2157 (1956).
3. N. A. Goryunova and S. I. Radautsan, Zh. Tekhn. Fiz. 28: 2917 (1958).
4. S. I. Radautsan, Investigation of Some Solid Solutions of Semiconductors Based on Indium Arsenide (dissertation for candidate's degree), Leningrad (1958).
5. J. C. Woolley, B. R. Pamplin, and J. A. Evans, J. Phys. Chem. Solids 19: 147 (1961).
6. D. N. Nasledov and I. A. Feltyn'sh, Fiz. Tverd. Tela 1: 565 (1959); 2: 823 (1960).
7. N. A. Goryunova and S. I. Radautsan, this volume, p. 1.
8. D. B. Gasson, P. J. Holmes, J. C. Jennings, J. E. Parrot, and A. W. Penn, Proc. Internatl. Conf. Semicond. Phys., held in Prague, 1960 (publ. Prague 1961), p. 1032.
9. S. I. Radautsan, I. A. Madan, and R. A. Ivanova, Izv. Moldavsk. Filiala Akad. Nauk SSSR, No. 3(69): 107 (1960).
10. S. I. Radautsan, Zh. Neorgan. Khim. 4: 1121 (1959).
11. I. I. Kozhina, S. S. Tolkachev, A. S. Borshchevskii, and N. A. Goryunova, Vestn. Leningr. Gos. Univ. No. 4: 122 (1962).
12. J. C. Woolley and B. A. Smith, Proc. Phys. Soc. (London) 72: 867 (1958).
13. M. S. Mirgalovskaya and E. V. Skudnova, Izv. Akad. Nauk SSSR, Otd. Tekhn. Nauk Met. i Toplivo, No. 4: 148 (1959).
14. N. A. Gorynova, Chemistry of Semiconductors, Izd. LGU (1963).
15. Anonymous, Mining J. 254: 133 (1960).
16. A. S. Borshchevskii and D. N. Tret'yakov, Fiz. Tverd. Tela 1: 1483 (1959).
17. H. Hahn and W. Klingler, Z. Anorg. Chem. 259: 135 (1949).
18. H. Welker and H. Weiss, Solid State Phys. 3: 1-78 (1956); see also Russian translation in: New Semiconducting Materials, IL (1958), p. 9.

ESTIMATE OF THE CHANGE IN THE CARRIER MOBILITY
WHEN A CONDUCTOR MELTS

A. R. Regel'

There are practically no unambiguous and exact experimental data on the carrier mobility in liquid conductors [1].

In view of this, it is useful to consider the possibility of making approximate estimates of the carrier mobility. We shall consider one of the variants of obtaining such estimates: the case when the electrical conductivity (σ_0) and the Hall coefficient (R_0) are known from experiment. We shall consider what conclusions we may draw about the carrier mobility from these experimental data if we use the standard phenomenological theory.

In the general case of the presence of n- and p-type conduction, as well as several groups of "light" and "heavy" electrons and holes, σ and R are given by:

$$\sigma = e \cdot \sum_{i=1}^{N} n_i u_i, \ R \cdot \sigma^2 = A \cdot \sum e_i n_i u_i^2, \tag{1}$$

where A is a coefficient governed by the nature of the relationship of the mean free path of a carrier with its kinetic energy. In the case considered here, the value of this coefficient ranges from 1 to 1.93. For simplicity, we shall assume that it is equal to 1. The other symbols in Eq. (1) have the following meaning: e_i is the elementary charge of positive or negative sign; n_i the density of carriers of given type; u_i the mobility of carriers of this type.

To analyze the possible errors in the calculation of the mobility, we shall consider the case of two types of carrier of different sign. We shall introduce the notation: $b = u_- / u_+$; $x = n_- / n_+$; and x_0, which satisfies the condition

$$x_0 \cdot b^2 = 1. \tag{2}$$

From the expressions in Eq. (1), we can easily calculate the relationship between the true carrier mobility (say, electron mobility, if the Hall coefficient is negative) and the effective mobility calculated on the assumption of the presence of carriers of one sign and therefore taken to be equal to $R\sigma$. It is readily seen that the effective mobility will always be smaller than the true mobility. The degree of the departure of the effective mobility from the true value will be determined by the value of b, and, what is more important, the degree of approach to the critical relationship $b^2 x_0 = 1$, i. e., to the condition

$$n_- u_-^2 = n_+ u_+^2. \tag{3}$$

The figure shows the results of such calculations for the case of dominant n-type conduction (negative Hall coefficient). The figure applies also to p-type conduction if b is replaced with $1/b$ and ($\log_{10}x - \log_{10}x_0$) is replaced by ($\log_{10}x_0 - \log_{10}x$). From the figure, it is evident that the effective mobility given by $Au_{eff} = R\sigma$ is

TABLE

Substance	σ_{sol}, ohm$^{-1}\cdot$cm^{-1}	R_{sol}, cm^3/C	$u_{sol} = R_{sol}\sigma_{sol}$, cm$^2\cdot$V$^{-1}\cdot$sec^{-1}	σ_{liq}, ohm$^{-1}\cdot$cm^{-1}	R_{liq}, cm^3/C	$u_{liq} = R_{liq}\sigma_{liq}$, cm$^2\cdot$V$^{-1}\cdot$sec^{-1}	u_{liq}/u_{sol}	References
K – Na	$2.5\cdot10^4$	$-4.2\cdot10^{-4}$	-10	$2\cdot10^4$	$-4.4\cdot10^{-4}$	-10	1	2,11
Rb	$8\cdot10^4$	$-5.0\cdot10^{-5}$	-4	$1.3\cdot10^5$	$-4.2\cdot10^{-5}$	-5	1.2	3,11
Hg	$5\cdot10^4$	$-8.7 \div 10.1\cdot10^{-5}$	-5	$1.06\cdot10^4$	$-7.5\cdot10^{-5}$	$-0,8$	0.2	4,11
Sb	$7\cdot10^3$	$+7\cdot10^{-3}$	$+50$	$1.05\cdot10^4$	$+2\cdot10^{-3}$	$\div20$	0.4	5,11
Te	$1.4\cdot10^2$	$+0.2$	$+30$	$2\cdot10^3$	$+2\cdot10^{-2}$	$+40$	1.3	1,6,10[*]
Bi$_2$Te$_3$	$2.6\cdot10^3$	$-7.5\cdot10^{-3}$	-20	$6.2\cdot10^3$	$-2.7\cdot10^{-3}$	-17	1	7
InSb with impurities	$2.5\cdot10^3$	-1.6	$-4\cdot10^3$	10^4	$\div5\cdot10^{-3}$	$\div50$	0.2	5,8[†]
InSb stoich.	$2\cdot10^3$	-4.0	$-8\cdot10^3$	$1.2\cdot10^4$	$-6\cdot10^{-3}$	-70	0.01	5,9[‡]

Note: a) the signs "$-$" and "$+$" denote, respectively, n– and p-type conduction;

b) the subscripts "sol" and "liq," denote, respectively, the solid and liquid states.

[*]$b = u_+/u_- \approx 1$ for Te at $T \geqslant 200°$C.

[†]$b = u_+/u_- \approx 20$ for InSb at $20°$C.

[‡]$b = u_+/u_- \approx 65$ for InSb at $20°$C.

109

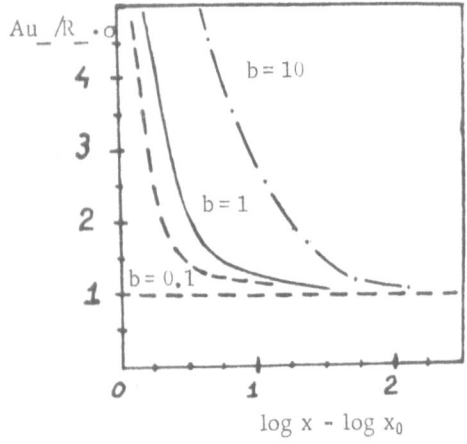

always less than the true x mobility, but it differs considerably from the true value only under conditions close to the critical condition $b^2 x_0 = 1$. If the carrier density differs from the critical value by a factor of 2-3, then for $b = 1$ the error in the calculated mobility is less than a factor of 2. Our estimates lead to the conclusion that a calculation of the carrier mobility using the approximate formula $u = R\sigma$ should, in most cases, give a good estimate of the true value of the mobility.

The table lists the experimental data on the electrical conductivity and Hall coefficient of several substances in the liquid and solid states.

Analysis of these data shows that in the majority of cases melting alters little the carrier mobility. In most cases, the change in the mobility does not exceed ±50%. N-type InSb, whose electron mobility in the molten state is approximately a hundred times smaller than in the solid state, is the only important exception. This sharp change in the mobility may not be real if InSb satisfies quite well the condition of Eq. (3): $n_- u_-^2 = n_+ u_+^2$. It is also possible that on melting the energy band structure of InSb may change, leading to, say, a sharp increase of the effective electron mass in the molten state.

It is interesting to note that even in those cases when the nature of the binding alters considerably on melting, the carrier mobility does not change greatly, except in the case of n-type InSb.

The values listed in the table were obtained by linear extrapolation of the experimental data to the melting point. For the substances listed, the carrier mobility in the molten state lies in the region 1-100 $cm^2 \cdot V^{-1} \cdot sec^{-1}$.

LITERATURE CITED

1. A. F. Ioffe and A. R. Regel', Progr. in Semiconductors, London 4: 239-291 (1960).
2. A. Kikoin and I. Fakidov, J. Phys. 71: 393 (1931).
3. I. Fakidov, Dokl. Akad. Nauk SSSR 63: 123 (1948).
4. P. W. Kendall and N. E. Cusack, Phil. Mag. 5: 100 (1960); 6: 419 (1961).
5. G. Busch and O. Vogt, Helv. Phys. Acta 27: 241 (1954).
6. E. Epstein, H. Fritzsche, and K. Lark-Horovitz, Phys. Rev. 107: 412 (1957).
7. V. P. Zhuze and A. R. Regel', Proc. Internatl. Conf. Semicond. Phys., held in Prague, 1960 (Prague 1961), p. 929.
8. S. S. Shalyt, Electrical Properties of Semiconductors.
9. H. Welker and H. Weiss, Solid State Phys. 3 (1956).
10. T. Fucuroi, RYTU 4: 353 (1952).
11. Ya. G. Dorfman and S. É. Frish (eds.), Collection of Physical Constants, ONTI (1937).

SOME TERNARY COMPOUNDS OF THE $A^I B_2^{IV} C_3^V$ TYPE

(Brief Communication)

V. I. Sokolova and E. V. Tsvetkova

A large number of ternary compounds with the zinc blende structure or with the similar chalcopyrite structure has been described [1-12]. These substances are semiconductors and belong, by their nature, to a large group of quite well-known substances which are members of the crystallochemical group of diamond. Since the number of new semiconducting materials in this group continuously increases (particularly by increasing the number of components), it has become necessary to classify these compounds. Goodman [2] made such an attempt by substituting one of the elements in a binary compound with its nearest neighbors in the periodic system. As a result of this, Goodman concluded that a very large number of compounds have the diamond-like structure.

Goryunova [3] suggested the following two rules in dealing with multicomponent tetrahedral phases: 1) "maximum valence," and 2) four electrons per atom. Using these rules, Goryunova showed analytically and graphically that the number of substances is not infinite but can be reduced to definite types. There are, for example, five types of ternary phase of interest in semiconductor technology:

TABLE 1

Type	Number of known compounds (according to published data)
$A^I B_2^{IV} C_3^V$	2 [1, 11]
$A_2^I B^{IV} C_3^{VI}$	8 [2]
$A^{II} B^{IV} C_2^V$	8 [4, 5, 9, 10]
$A^I B^{III} C_2^{VI}$	20 [6, 8]
$A_3^I B^V C_4^{VI}$	4 [2]

It is evident from Table 1 that the preparation of these compounds has not yet been investigated much. The least known is the $A^I B_2^{IV} C_3^V$ group. The first data on this group were reported by Goryunova and Sokolova [1].

In the work reported here, an attempt was made to prepare substances of this group. Table 2 lists all possible combinations in that group.

We investigated the combinations of elements which are framed in Table 2. All the substances were prepared by the method of fusing together the components. X-ray diffraction and microstructure analyses were used to check the phase composition.

During this study it was discovered that the combinations containing lead and bismuth gave off free bismuth. The alloys of tin with antimony or arsenic revealed the presence of the NaCl structure, corresponding to the formation of the binary compounds SnSb and SnAs, respectively. Alloys based on germanium and containing antimony or arsenic usually gave off pure germanium. In all these cases, two or more phases were present.

TABLE 2

Cu	Ag	Au
$CuSi_2P_3$	$AgSi_2P_3$	$AuSi_2P_3$
$CuGe_2P_3$	$AgGe_2P_3$	$AuGe_2P_3$
$CuSn_2P_3$	$AgSn_2P_3$	$AuSn_2P_3$
$CuPb_2P_3$	$AgPb_2P_3$	$AuPb_2P_3$
$CuSi_2As_3$	$AgSi_2As_3$	$AuSi_2As_3$
$CuGeAs_3$	$AgGe_2As_3$	$AuGe_3As_3$
$CuSn_2As_3$	$AgSn_2As_3$	$AuSn_2As_3$
$CuPb_2As_3$	$AgPb_2As_3$	$AuPb_2S_3$
$CuSi_2Sb_3$	$AgSi_2Sb_3$	$AuSi_2Sb_3$
$CuGe_2Sb_3$	$AgGe_2Sb_3$	$AuGe_2Sb_3$
$CuSn_2Sb_3$	$AgSe_2Sb_3$	$AuSn_2Sb_3$
$CuPb_2Sb_3$	$AgPb_2Sb_3$	$AuPb_2Sb_3$
$CuSi_2Bi_3$	$AgSi_2Bi_3$	$AuSi_2Bi_3$
$CuGe_2Bi_3$	$AgGe_2Bi_3$	$AuGe_2Bi_3$
$CuSn_2Bi_3$	$AgSe_2Bi_3$	$AuSn_2Bi_3$
$CuPb_2Bi_3$	$AgPb_2Bi_3$	$AuPb_2Bi_3$

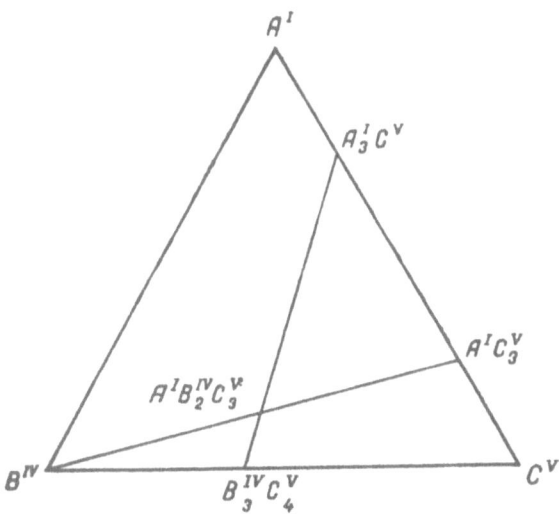

Fig. 1. Gibbs triangle of the Cu−Ge−P system.

The reactions with silicon requiring high temperatures were not completed under the conditions used by us. Only the phase $CuSi_2P_3$, crystallizing with the ZnS structure and the parameter $a = 5.25$ kXU, was obtained. Our initial attempts to indroduce gold into compounds with a tetrahedral structure were not successful. Therefore, we restricted ourselves to a single attempt (Table 2) which showed that gold reacts only partly with germanium and arsenic.

Alloys containing phosphorus consisted of a single phase in two cases: $CuGe_2P_3$ and $AgGe_3P_3$. Two other compositions, $CuSn_2P_3$ and $AgSn_2P_3$, showed quite a complex phase composition in their diffraction patterns. The main phase of $CuSn_2P_3$ crystallized in the ZnS structure, while the main phase of $AgSn_2P_3$ had a complex structure.

The substances $CuGe_2P_3$ and $AgGe_2P_3$ were subjected to a more detailed investigation.

The substance represented by $CuGe_2P_3$ crystallized in the zinc blende structure with the parameter $a = 5.35_8$ kXU. The microstructure analysis showed the presence of 3-5% of a second phase. The value of the microhardness of the main phase was 850 ± 20 kg/mm^2. The microhardness of the second phase could not be measured. The thermal analysis, carried out between 200 and 1300°C, also showed the presence of two transformations: 1) at 800°C; 2) at 759°C.

To obtain more information on $CuGe_2P_3$, we made an attempt to study its homogeneous region. We started from general considerations regarding the formation of ternary tetrahedral phases (Fig. 1). It is evident from Fig. 1 that if we consider the possibility of the formation of nondefect tetrahedral ternary phases of variable composition, we find that they can be formed only along the tie-line $B^{IV}-A^IB^{IV}C_3^V$. Defect phases, for which the average number of electrons (four) should be calculated per lattice site and not per atom, can be formed along the tie-line $B_3^{IV}C_4^V-A^IB_2^{IV}C_3^V$. In the composition triangle Cu−Ge−P, we found along the tie-line $CuGe_2P_3-$Ge a region of mainly single-phase substances from $CuGe_2P_3$ to $CuGe_5P_3$, the lattice parameter of which varied in accordance with Vegard's law. A phase with the zinc blende structure was found to exist over a wide region along the tie-line $CuGe_2P_3-Ge_3P_4$.

$AgGe_2P_3$, unlike $CuGe_2P_3$, does not crystallize in the zinc blende structure. The lattice of $AgGe_2P_3$ belongs to the cubic system, its unit cell being a body-centered cube. $AgGe_2P_3$ samples were dense and consisted of a single phase. The value of the microhardness was 730 ± 20 kg/mm^2. As in the case of $CuGe_2P_3$, the microhardness was measured using etched and unetched sections. The etchant was a mixture of nitric and acetic acids.

The thermal analysis, carried out between 200 and 1300°C, established one transformation in $AgGe_2P_3$ at 742°C.

Thus, the $A^IB_2^{IV}C_3^V$ type of ternary system was found to consist of a relatively small number of substances. However, the members of this group which do exist exhibit interesting properties. Thus, $CuGe_2P_3$ dissolves a large (of the order of 30 mol. %) amount of germanium, a property not exhibited by any other known binary or more complex substance.

All alloys in the homogeneous region, i. e., alloys from $CuGe_2P_3$ to $CuGe_5P_3$ exhibited one system of lines, representing the ZnS structure, and these lines were sharp. The period varied from $a_W = 3.58$ kXU for $CuGe_2P_3$ to $a_W = 5.48$ kXU for $CuGe_5P_3$. The large difference between the periods of the Ge and $CuGe_2P_3$ lattice, considerably greater than the experimental error (for Ge, $a_W = 5.65$ kXU), showed clearly that solid solutions were formed in alloys of these compounds.

We discovered also the formation of defect homogeneous phases along the tie-line $CuGe_2P_3-Ge_3P_4$.

The nature of the morphotropic transition from $CuGe_2P_3$ to $AgGe_2P_3$ and from $CuGe_2P_3$ to $AgSn_2P_3$ when the atoms in one group of compounds were substituted was very interesting: this transition was accompanied by the formation of a substance with a new structure, which had no analogs in the substances so far investigated along the morphotropic transition boundaries of systems having the zinc blende structure.

In conclusion, the authors thank N. A. Goryunova for discussing the results and for her valuable advice.

LITERATURE CITED

1. N. A. Goryunova and V. I. Sokolova, Izv. Moldavsk. Filiala Akad. Nauk SSSR, No. 3(69) (1961).
2. C. H. L. Goodman, J. Phys. Chem. Solids 6: 306 (1958).
3. N. A. Goryunova, Vestn. Leningradsk. Univ. 10: 62, 112 (1961).
4. C. H. L. Goodman, Nature (London) 179: 828 (1957).
5. O. G. Folberth and H. Pfister, Acta Cryst. 13: 199 (1960).
6. H. Hahn, G. Frank, W. Klingler, A. D. Meyer, and G. Störger, Z. Anorg. Allgem. Chem. 271: 153(1953).
7. I. G. Austin, C. H. L. Goodman, and A. E. Pengelly, Nature (London) 178: 433 (1956).
8. C. H. L. Goodman, I. G. Austin, and A. E. Pengelly, J. Electrochem. Soc. 103: 609 (1956).
9. H. Pfister, Acta Cryst. 11: 221 (1958).
10. O. G. Folberth and H. Pfister, Halbleiter and Phosphore (1958), p. 474.
11. O. G. Folberth and H. Pfister, Acta Cryst. 14: 325 (1961).
12. N. A. Goryunova, A. A. Vaipolin, and Tseng Ping-hsi, Papers presented at the Nineteenth Scientific Conference of LISI, Physics and Chemistry (1961).

THE POSSIBILITY OF THE FORMATION OF SOLID SOLUTIONS IN SOME SYSTEMS BASED ON $A^{III}B^{V}$

(Brief Communication)

L. V. Kradinova

The mobility of carriers in semiconducting compounds of the $A^{III}B^{V}$ type increases with increase of the atomic weight of the component elements. Thus, InSb, which is the last member of the homologous series $A^{III}B^{V}$, is one of the most interesting semiconductors, having the maximum value of the mobility for this series.

Tha aim of the present work was to prepare samples with the heaviest elements (Tl or Bi), using the $A^{III}B^{V}$ series as the base, in order to increase the mobility. At present, no known binary compound with the ZnS structure contains Tl or Bi. Only Hahn [1] has reported ternary compounds with the chalcopyrite structure, in which Tl exhibited its maximum valence and was tetrahedrally surrounded. Our experiments on the synthesis of these compounds, and the experiments of other authors [2] did not confirm Hahn's findings. However, this does not mean that solid solutions including Tl cannot be formed, because there are known cases of solid solutions based on hypothetical compounds [3]. An attempt was made to introduce thallium or bismuth either by dissolving them in $A^{III}B^{V}$ substances with the ZnS structure, analogous to the hypothetical compounds based on Tl or Bi (GaBi, etc.), or by introducing Tl or Bi as elements into $A^{III}B^{V}$. To investigate the samples obtained, x-ray diffraction structure analysis, microstructure analysis, or microhardness measurements were carried out.

The results showed that the expected solid solutions (or compounds) were not obtained in the range of concentrations defined by the technique used. This may be explained by the very strong "metallization" of the bonds on transition from In to Tl and from Sb to Bi. However, we may conclude that it should be possible to introduce Tl or Bi into more complex compounds with the zinc blende structure. This conclusion is supported to a certain extent by Prochukhan's tests [3], in which tin was introduced into a quaternary alloy based on antimonides with the ZnS structure, while the corresponding ternary alloy with tin was not obtained.

LITERATURE CITED

1. H. Hahn and W. Klingler, Z. Anorg. Allgem. Chem. Vol. 271 (1953).
2. I. Black and E. Banks, Proc. Internatl. Conf. Semicond. Phys. held in Prague, 1960 (Publ. Prague 1961).
3. N. A. Goryunova and V. D. Prochukhan, Fiz. Tverd. Tela 2: 176 (1960).

INDEX